AQA Science

Science B: Science in Context

Revision Guide

New GCSE

James Hayward
Jo Locke
Nicky Thomas
Andrea Johnson

Series Editor
Lawrie Ryan

D1581597

Nelson Thornes

GCSE Science B: Science in Context

Welcome to AQA GCSE Science!

Key points

At the start of each topic are the important points that you must remember.

This book has been written for you by the people who will be marking your exams, very experienced teachers and subject experts. It covers everything you need to revise for your exams and is packed full of features to help you achieve the very best that you can.

Key words are highlighted in the text and are shown **like this**. A glossary of these terms can be found within Kerboodle, www.kerboodle.com

▐▐▐➡ *These questions check that you understand what you're learning as you go along. The answers are all at the back of the book.*

Many diagrams are as important for you to learn as the text, so make sure you revise them carefully.

Anything in the Higher boxes must be learned by those sitting the Higher Tier exam. If you'll be sitting the Foundation Tier, these boxes can be missed out.

The same is true for any other places that are marked [H].

Higher

AQA Examiner's tip

AQA Examiner's tips are hints from the examiners who will mark your exams, giving you important advice on things to remember and what to watch out for.

Bump up your grade

How you can improve your grade – this feature shows you where additional marks can be gained.

Maths skills

This feature highlights the maths skills that you will need for your Science exams with short, visual explanations.

And at the end of each chapter you will find:

End of chapter questions

These questions will test you on what you have learned throughout the whole chapter, helping you to work out what you have understood and where you need to go back and revise.

And at the end of each unit you will find:

AQA Examination-style questions

These questions are examples of the types of questions you will answer in your actual GCSE, so you can get lots of practice during your course.

You can find answers to the End of chapter and AQA Examination-style questions at the back of the book.

Student Book
pages 4–5

1.1 Observing our solar system

- **Telescopes** are used to observe **stars** and other distant objects as very little light from them reaches us. Telescopes collect and focus their light to make a brighter, bigger image.
- Ground-based telescopes can be updated, maintained and visited easily. However, our atmosphere can spoil the images. Space-based telescopes take clear images because there is no light pollution or atmosphere to interfere, but they are expensive to run and hard to mend or visit. Telescopes can provide evidence of changes taking place in the universe.

> **1** *Write down one advantage and one disadvantage of a space-based telescope.*

- Some stars and planets emit radio waves, and newly formed stars and dust clouds emit microwaves. Ground-based telescopes detect visible light, **radio waves** and **microwaves**, which pass through our atmosphere.
- Space-based telescopes detect visible light, **infrared radiation** and **ultraviolet radiation** from stars. They also detect **X-rays** and **gamma rays** from neutron stars and from near black holes.

> **2** *What types of radiation do telescopes detect?*

- Space probes travel through the solar system and may land on other **planets**. They gather data and send it back to Earth.

Key words: telescope, star, radio wave, microwave, infrared radiation, ultraviolet radiation, X-ray, gamma ray, planet

Key points

- Telescopes are used to detect visible light and other electromagnetic radiations coming from distant objects.
- Telescopes can be based on the ground or in space.
- Different telescopes see different types of electromagnetic radiation, allowing us to see different objects in space.
- These observations provide evidence of changes in the universe.

Examiner's tip

Read the question carefully before you answer it – it could be about ground-based telescopes or space-based telescopes.

Student Book
pages 6–7

1.2 How the universe began

- Waves from objects moving away from an observer appear to stretch. This is the **Doppler effect**. Red light has a longer wavelength than blue light. When a star or galaxy moves away from us, its light appears redder than expected because its wavelength is stretched. This effect is called the **red-shift**.

Key points

- The observed wavelength and frequency of a moving light source change (the Doppler effect).
- The red-shift is evidence that the universe is expanding.
- The universe began as a huge explosion called the Big Bang.
- The cosmic background radiation is evidence for the theory that the Big Bang took place.

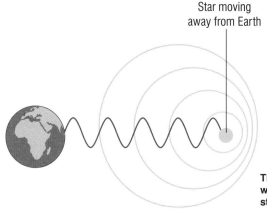

Star moving away from Earth

The Doppler effect means that waves are stretched apart if a star moves away from the Earth

> **1** *What is the red-shift?*

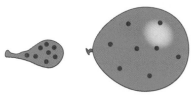

As the balloon gets bigger the spots spread apart, just as galaxies move away from each other as the universe expands

- Galaxies further away from us have a bigger red-shift so they must be moving away faster than closer galaxies. Galaxies in all directions move away from Earth, and all other galaxies, so the whole universe must be expanding.
- Many scientists believe that all matter and energy in the universe were once squashed in one tiny place. About 14 billion years ago a huge expansion started called the **Big Bang**. Matter was flung outwards and the universe is still expanding today.
- **Cosmic microwave background radiation** is a weak microwave signal arriving at Earth from all directions. Scientists believe this is the remains of radiation created during the Big Bang.

▐▐▶ **2** *Write down two pieces of evidence that the universe began from a very small point as a huge explosion.*

Key words: Doppler effect, red-shift, Big Bang, cosmic microwave background radiation

Student Book
pages 8–9

Key points

- Our models of the solar system have changed over centuries.
- The Earth orbits the Sun, which is at the centre of our solar system which is in the Milky Way galaxy.
- It is likely there are other places in the universe that can support life.

1.3 Our place in the universe

- Stars are balls of immensely hot, glowing gas. Planets follow a path called an orbit around a central star and are visible because they reflect the star's light.

▐▐▶ **1** *Write down one difference between a star and a planet.*

- Over 2000 years ago, some people believed the Earth, Moon and stars orbited a central fire, or that the Sun, Moon and planets orbited the Earth. About 500 years ago, people realised planets orbit the Sun in our **solar system**. The **Moon** orbits the Earth. A planet's **orbit** is an ellipse (squashed circle).

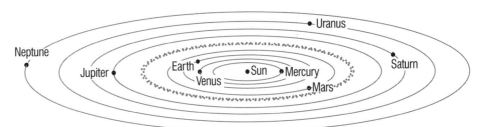

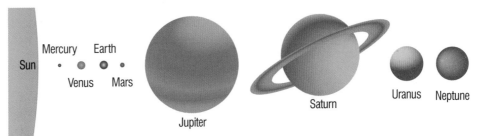

The modern model of the solar system, where eight planets orbit the Sun on ellipses

- There are hundreds of billions of stars in our **galaxy**, the Milky Way. In the universe, there are hundreds of billions of galaxies.
- There may be other planets like Earth in our galaxy, the right distance from a star for water to be a liquid. Liquid water is needed to sustain life.

▐▐▶ **2** *Why is it likely that there is life on other planets in the universe?*

Key words: solar system, Moon, orbit, galaxy

1.4 The Earth's structure

Student Book
pages 10–11

Key points

● The Earth was originally a ball of molten rock, which slowly cooled down to form a crust.

● Today, the Earth consists of a metallic core surrounded by a mantle, a thin crust and a gaseous atmosphere.

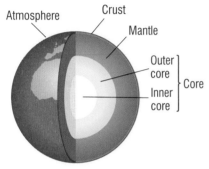

The Earth's layered structure

● The Earth separated into layers when it first formed and cooled from a ball of **molten** material. The heavier materials sank to the centre, and the lighter ones stayed nearer the surface.

● By studying the vibrations caused by earthquakes, geologists learn about the structure of the Earth. The outer **crust** we live on is quite thin. Beneath the crust is the **mantle**, which flows slowly. The centre of the Earth is called the **core** and is made of iron and nickel. The outer core is a molten liquid, the inner core is solid. Around the outside is a relatively thin layer of gases called the **atmosphere**.

▥▶ **1** *How many different layers make up the Earth's structure? What are their names?*

Key words: molten, crust, mantle, core, atmosphere

1.5 Changes in the Earth's surface

Student Book
pages 12–13

Key points

● Tectonic plates are made up of the Earth's crust and upper part of its mantle.

● Heat from the Earth's core causes convection currents in the mantle which cause tectonic plates to move.

● Earthquakes and volcanoes occur mainly at boundaries between tectonic plates and are difficult to predict accurately.

Bump up your grade

Just linking tectonic plates to volcanoes and earthquakes is a basic skill. To get more marks, describe **how** the different plate boundaries cause earthquakes and volcanoes.

● Nuclear reactions deep inside the Earth heat up the mantle, reducing its density. This causes the hottest parts of the mantle to rise up by convection. These **convection currents** have cracked the crust into huge pieces we call **tectonic plates**. The tectonic plates move very slowly around the surface of the Earth.

▥▶ **1** *What causes convection currents in the Earth's mantle?*

● When one plate slips under another one, it starts to melt. The molten rock sometimes rises back to the Earth's surface, forming a **volcano**.

● When plates tend to slide past each other, very strong forces build up between them. This can make them suddenly slip past each other, causing an **earthquake**.

● Scientists cannot make accurate predictions of these events, but some progress is being made.

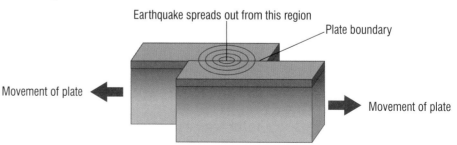

Earthquakes are caused when plates grind against each other

Key words: convection current, tectonic plate, volcano, earthquake

Student Book
pages 14–15

1.6 The Earth's changing atmosphere

Key points

- The Earth's early atmosphere was produced by volcanoes, and was mainly carbon dioxide.

- There may also have been water vapour and small amounts of methane, hydrogen and ammonia in the early atmosphere [H].

- As the water vapour cooled, the first oceans formed and early plants and algae began to evolve.

- Early microorganisms used carbon dioxide in the atmosphere for photosynthesis. This made oxygen gas.

- The Earth's atmosphere has changed a lot since our planet was formed. Originally it did not support living organisms like us. These changes happened over millions of years.

- Earth's first atmosphere came from gases released by volcanoes. It was mainly **carbon dioxide** with little or no oxygen.

There might also have been **water vapour**, and a small proportion of ammonia, hydrogen and methane. This mixture of gases was shown to be capable of making amino acids – the molecules of life – in experiments in the 1950s carried out by Miller and Urey.

- The water vapour cooled down and condensed into oceans. The first plants and algae to evolve started producing oxygen through **photosynthesis**.

▭▶ **1** *Why did carbon dioxide levels start going down once plants had evolved?*

Key words: photosynthesis

Student Book
pages 16–17

1.7 Maintaining our atmosphere

Key points

- The Earth's atmosphere contains mainly nitrogen and oxygen, with small amounts of carbon dioxide, water vapour and argon.

- Short-wave radiation (light energy) from the Sun passes through the atmosphere to warm the Earth.

- Today, Earth's atmosphere is mainly nitrogen and oxygen, with a tiny amount of carbon dioxide and water vapour. There are also trace amounts of gases such as argon.

- **Greenhouse gases**, such as carbon dioxide and methane, trap energy in the Earth's atmosphere. Energy reaches Earth from the Sun by **short-wave radiation** and **ultraviolet radiation**. When the Earth warms up, it releases energy into space as **long-wave radiation (infrared)**, which is absorbed by greenhouse gases. This can be useful because it makes Earth warm enough for us to live on. However, too much greenhouse gas in the atmosphere can cause global warming and extreme changes to weather patterns. This is called **climate change**.

▭▶ **1** *Why can't all long-wave radiation escape from Earth into space?*

AQA Examiner's tip

Do not make the mistake of saying that greenhouse gases **reflect** energy back to Earth. They don't! Greenhouse gases **absorb** energy, keeping the atmosphere warm.

Key words: greenhouse gas, climate change

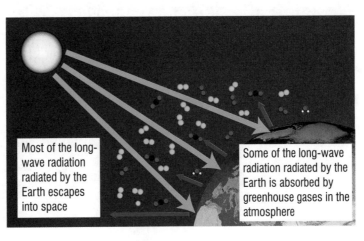

Most of the long-wave radiation radiated by the Earth escapes into space

Some of the long-wave radiation radiated by the Earth is absorbed by greenhouse gases in the atmosphere

Greenhouse gases trap energy inside Earth's atmosphere

1 Explain why we use telescopes that can detect different types of electromagnetic radiation.

2 Why do space-based telescopes produce clearer images than ground-based telescopes?

3 What evidence is there that the universe is expanding?

4 What is the Doppler effect?

5 Draw a flow chart to show how nuclear reactions deep inside the Earth can cause volcanoes.

6 Make a table to summarise the different parts of the Earth's structure. Use the headings: 'Name of layer', 'Location', 'Description'.

7 Why was Earth's early atmosphere mainly carbon dioxide?

8 How did Earth's oceans form?

9 Explain why carbon dioxide in the atmosphere is both necessary for life and potentially dangerous to life.

10 Why can greenhouse gases in the atmosphere cause the Earth's temperature to rise?

11 Describe how Miller and Urey's experiments suggested the way life might have started on Earth.

[H]

Chapter checklist	✔	✔	✔
Tick when you have:			
reviewed it after your lesson	✔	☐	☐
revised once – some questions right	✔	✔	☐
revised twice – all questions right	✔	✔	✔
Move on to another topic when you have all three ticks			

	✔	✔	✔
Observing our solar system	☐	☐	☐
How the universe began	☐	☐	☐
Our place in the universe	☐	☐	☐
The Earth's structure	☐	☐	☐
Changes in the Earth's surface	☐	☐	☐
The Earth's changing atmosphere	☐	☐	☐
Maintaining our atmosphere	☐	☐	☐

2.1 Building blocks of new products

Student Book
pages 20–21

Key points

- An element is made of only one type of atom.
- Compounds contain more than one type of atom, bonded together.
- Mixtures contain more than one type of element or compound, not bonded together.

Examiner's tip

Learning the symbols of the most common elements will speed up your ability to identify them. Learn the names and symbols of the first 20 elements to get a head start in exams.

- All materials are made of **atoms**, and can be divided up into three groups: **elements**, **compounds** and **mixtures**.
- Each different type of atom is called an element. If a material is an element, all its atoms are the same type. If the material is in the **periodic table**, then it's an element. The diagram shows the metallic element gold (chemical symbol: Au).
- Compounds are substances where elements have bonded (chemically joined) to each other. This is easy in the example below (it is water). If symbols are used, you would see the symbols for more than one element – in this case, H_2O.
- Mixtures are combinations of different compounds and/or elements. They have not been bonded together. There are two different types of atom mixed with the water molecules in the example below.

▐▶ **1** *How many types of atom are in the diagram of a mixture below?*

An element, gold

A compound, water

A mixture

- The periodic table is a list of all the elements. There are about 100 different elements. Numbers on the periodic table give us more information about each element.

Key words: atom, element, compound, mixture, periodic table

2.2 Inside atoms

Student Book
pages 22–23

Key points

- Protons are positive, neutrons are neutral and electrons are negative.
- The number of protons in an atom equals the number of electrons in that atom.
- Protons and neutrons are in the nucleus of an atom and have a relative mass of 1.
- Electrons have a very tiny mass and orbit the nucleus.

- An atom is the basic building block of all materials. The centre of an atom is called the **nucleus**. It is made of two types of particle: **protons** and **neutrons**. The rest of an atom is a cloud of **electrons**. The electrons **orbit** the nucleus.
- The particles in atoms have different charges and masses from each other:

Particle	Where is it?	Charge	Mass
Proton	In the nucleus	+1	1
Neutron	In the nucleus	0	1
Electron	Orbiting the nucleus	−1	Almost zero

- The overall charge on an atom is zero, so atoms must have the same number of electrons as protons.
- The periodic table tells you the **atomic number** and **mass number** of each element. This lets you calculate the numbers of protons, neutrons and electrons in an atom.
- **atomic number** = number of protons (which equals the number of electrons)
- **mass number** = number of protons + number of neutrons

So, the number of neutrons = mass number – atomic number

▐▶ **1** *The atomic number of carbon is 6 and its atomic mass is 12. How many protons, neutrons and electrons does a carbon atom have?*

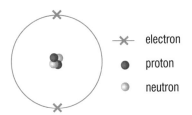

electron

proton

neutron

A helium atom

Key words: nucleus, proton, neutron, electron, atomic number, mass number

Student Book
pages 24–25

2.3 Different types of particles

Key points

- Atoms are the building blocks of all materials.
- A molecule is two or more atoms that are chemically bonded together.
- If an atom or molecule becomes positively or negatively charged, it is called an ion.

- Sometimes atoms are joined together with chemical bonds, and sometimes they have lost or gained electrons.
- **Molecules** are created when two or more atoms are bonded together. O_2 is a molecule because it contains two oxygen atoms.

$$O=O \qquad N\equiv N \qquad O=C=O$$

Oxygen   Nitrogen   Carbon dioxide

$$H-\overset{\displaystyle H}{\underset{\displaystyle H}{C}}-H$$

Methane

Some common molecules

▶ **1** *How many atoms are in each of the molecules in the diagram? Write them out using symbols.*

- An **Ion** is created when an atom loses or gains electrons. Because electrons are negatively charged, gaining them makes a negative ion. Losing electrons makes a positive ion.

Key words: molecule, ion

Student Book
pages 26–27

2.4 Making products with materials from the Earth

Key points

- Gold, marble, limestone and sulfur can all be used straight from the ground.
- Crude oil is a mixture of hydrocarbons. We can separate hydrocarbons by fractional distillation.
- Pure salt can be extracted from rock salt by adding water and filtering out the rock. We can get the salt from its solution by evaporating off the water.

- Materials we take from the environment can be divided into two groups: materials we can use straight from the ground, and materials we need to separate.

Materials we can use straight from the ground:

Material	Uses
Gold	Electronics, jewellery
Marble	Building
Limestone	Building, making other materials
Sulfur	Sulfuric acid, fertilisers, car batteries

Materials we need to separate (and how we do it):

Material	How it is separated	Uses
Crude oil (a mixture of different-sized **hydrocarbon** molecules)	**Fractional distillation**. Different-sized hydrocarbons will boil at different temperatures.	Fuels, plastics
Rock salt (a mixture of rock and salt)	Filtration and evaporation. The salt is soluble but the rock is not.	Flavouring food

- We take advantage of materials' physical properties when we separate them. Usually these are differences in boiling point or solubility.

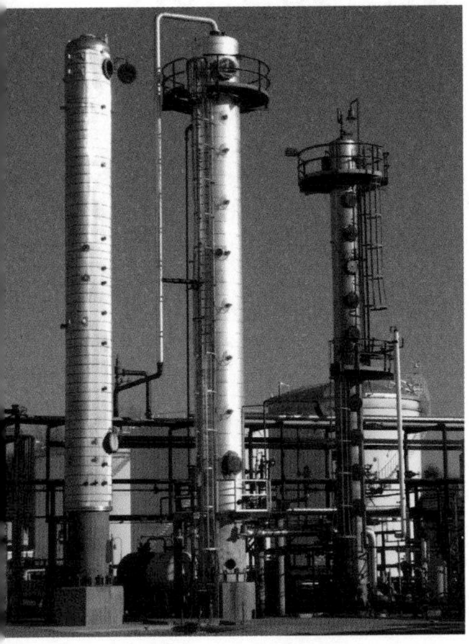

Fractional distillation separates crude oil into useful products in an oil refinery

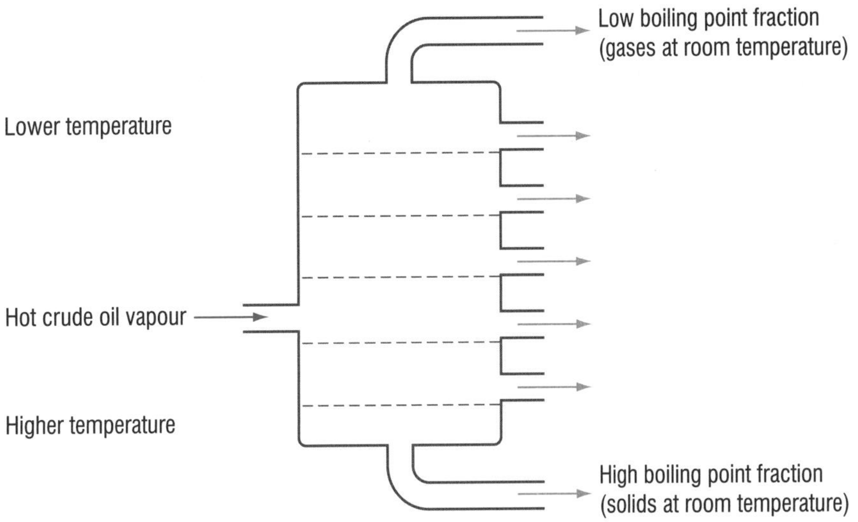

Different-sized hydrocarbons have different boiling points. This helps us separate them by fractional distillation.

1 *List two products that are made from crude oil.*

Key words: hydrocarbon, fractional distillation

Student Book
pages 28–29

2.5 Extracting metals for construction

Key points

- Reducing agents remove oxygen from ores.
- Reducing agents must be more reactive than the metal in the ore.
- Iron oxide and lead oxide are reduced by carbon and carbon monoxide.
- More reactive metals like aluminium are extracted by electrolysis.

- An **ore** is a rock from which it is economically worth while to extract a metal. Many ores are rich in metal oxide. Ores can be **reduced** to remove the oxygen.
- **Reducing agents** remove oxygen from oxides. The reducing agent must be more **reactive** than the metal for this to happen. A **reactivity series** is a list of chemicals in order of reactivity – this is helpful when choosing a reducing agent.
- To produce iron, carbon is added to iron oxide and air is blasted through the mixture. The carbon monoxide made reduces the iron oxide. A furnace heats it all to 1500°C and the following reactions happen:

carbon + oxygen ⟶ carbon dioxide $C + O_2 \longrightarrow CO_2$

carbon + carbon dioxide ⟶ carbon monoxide $C + CO_2 \longrightarrow 2CO$

carbon monoxide + iron oxide ⟶ carbon dioxide + iron $3CO + Fe_2O_3 \longrightarrow 3CO_2 + 2Fe$

- In the extraction of lead, carbon is added to lead oxide and the mixture is heated. Carbon and carbon monoxide both reduce the lead oxide.

carbon + lead oxide ⟶ carbon monoxide + lead $C + PbO \longrightarrow CO + Pb$

carbon monoxide + lead oxide ⟶ carbon dioxide + lead $CO + PbO \longrightarrow CO_2 + Pb$

1 *Name the reducing agents used in the extraction of iron and lead from their oxides.*

- **Electrolysis** is an expensive process used to extract more reactive metals like aluminium. Aluminium oxide is melted and an electrical current is passed through it. This separates the aluminium and oxygen.

Key words: ore, reducing agent, reactivity series, electrolysis

AQA Examiner's tip

At Higher Tier you may be asked to write balanced equations for the reactions involved in separating metals from their ores.

Student Book
pages 30–31

2.6 Products from the atmosphere

- The atmosphere is a mixture of nitrogen, oxygen, water vapour, a little carbon dioxide and some trace gases. To extract these gases, air is compressed and cooled.
- The temperature of the liquid ($-200\,°C$) is slowly raised. This causes the gases to boil off one by one because they have different boiling points. They can then be stored separately.

⟫ **1** *What are the similarities and differences between separating the hydrocarbons in crude oil and the gases in liquefied air?*

Gas	Uses	More details
Nitrogen	Making ammonia Freezing agent Food preservative	Ammonia is used to make **fertilisers** Liquid nitrogen boils at $-196\,°C$ Nitrogen gas is very unreactive
Helium	Balloons, airships Coolant in MRI machines	Helium is much less dense than air Liquid helium boils at $-269\,°C$
Argon	Filament light bulbs Lasers	Argon will not react with the metal filament Argon produces light in an electric field

Key words: fertiliser

Key points

- Gases are separated from the air by liquefying air and separating the liquids by fractional distillation. [H]
- Argon is used mainly in lighting. [H]
- Helium is used for cooling and for filling balloons and airships. [H]
- Nitrogen is mainly used to make ammonia. [H]

Student Book
pages 32–33

2.7 Exploiting the Earth's resources

- Products from mines and quarries are important to society. Removing ores from the ground also makes money and provides jobs for people.

There are problems with taking materials from the ground:

- **Mines** and **quarries** damage the appearance of the landscape.
- Dust and noise affect the nearby area.
- Mining and quarrying increase traffic, causing more air pollution.
- Processing ores produces carbon dioxide, adding to global warming.
- Toxic minerals released by waste from mining can poison soil and lakes.

Extracting gases from the air is less harmful than mining and quarrying, as air is so readily available. However, there are still some drawbacks:

- Liquefying air requires a lot of energy, so more fossil fuels are burned to provide it.
- Industrial plants that generate gas from the air take up land.
- Some products can be harmful if used without precautions, such as ammonia which is made from the gases taken from the air.

Resources have to be managed sustainably:

- **Sustainability** means using resources in a way that finds a balance between human and environmental needs.
- **Stakeholders** are consulted and long-term impact on local communities is considered.
- Ways to reduce the environmental impact of the operation are considered.

Phytomining is a way of reducing the impact of mines and quarries. Plants are used to remove minerals from the soil without harming the environment.

Key points

- We benefit from the products, services and jobs provided by taking materials from the Earth.
- The environment can be damaged by mining operations.
- Managing mining sustainably reduces long-term environmental damage.
- Plants can be used to phytomine areas, absorbing potentially toxic minerals from soil. [H]

⟫ **1** *What long-term effects might a mine have on a nearby town?*

Key words: mine, quarry, sustainability, phytomining

Student Book
pages 34–35

2.8 Changing materials to make new products

Key points

- During chemical reactions, the atoms in the reactants rearrange to form the products.
- The total mass of products will always be the same as the total mass of reactants.
- The mass of one reactant can be calculated if you know the mass of the products and any other reactants.

- The materials you start with in a chemical reaction are called **reactants**. The materials you end up with are called the **products**. Here is an example:

  methane + oxygen $\longrightarrow$ carbon dioxide + water

- Methane and oxygen are reactants, carbon dioxide and water are products.
- If 10 kg of methane and oxygen is used, then 10 kg of carbon dioxide and water will be produced. This is called the Law of the **Conservation of Mass**. All that changes is how the atoms are arranged. This is because the bonds between atoms break and different bonds are formed.

> **1** *Why is the total mass of reactants always the same as the total mass of products in a chemical reaction?*

> **2** *What are the reactants and products for the final stage of processing iron ore? (Turn back to page 9 and look at the word equations.)*

> **3** *If 446 tonnes of lead oxide and 12 tonnes of carbon are smelted, 44 tonnes of carbon dioxide will be produced. How much lead will be produced?*

What goes in must come out. The mass of products always equals the mass of reactants.

Maths skills

A chemist decomposes 10 g of calcium carbonate by heating it strongly:

calcium carbonate $\longrightarrow$ calcium oxide + carbon dioxide

If 5.6 g of calcium oxide is collected, there must have been 4.4 g of carbon dioxide given off. Why?

Bump up your grade

Descriptions such as 'reactants form products when the atoms are rearranged' can become more highly scoring explanations by adding, '… because the bonds between them are broken and then re-made'.

AQA Examiner's tip

The **total** mass of reactants equals the **total** mass of products. Make sure you've taken all the products and reactants into account when you calculate masses.

Key words: reactant, product, conservation of mass

Student Book
pages 36–37

Key points

- Chemical equations are used to show which chemicals react and what is produced.
- Balanced chemical equations help chemists calculate how much raw material they need. [H]

AQA *Examiner's tip*

If an exam question asks you to write a symbol equation, make sure you balance it.

Maths skills

Remember that writing a number in front of a formula will affect the numbers of **each** atom within it. For example, $2H_2SO_4$ contains **4** hydrogen atoms, **2** sulfur atoms and **8** oxygen atoms.

2.9 Using equations

- In a **chemical formula**, numbers tell us how many atoms are present. Small numbers after an element tell us how many atoms there are of that element.

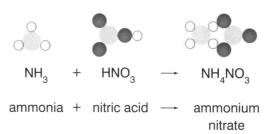

$$NH_3 \quad + \quad HNO_3 \quad \longrightarrow \quad NH_4NO_3$$

ammonia + nitric acid ⟶ ammonium nitrate

Making fertiliser

- In the reaction above, you can see ammonia is made of one nitrogen atom and three hydrogen atoms.

⟼ **1** *What are the names of the different elements in ammonium nitrate?*

A **balanced equation** shows how much of each reactant is needed. This is shown in equations by writing numbers in front of a formula. Look at the example below:

$$3\,H_2 \quad + \quad N_2 \quad \longrightarrow \quad 2\,NH_3$$

Hydrogen and nitrogen react to produce ammonia

Three times as many hydrogen molecules as nitrogen molecules are needed.

To balance an equation, count the numbers of each type of atom in the reactants and products. Adjust the numbers of molecules until the number of each type of atom in the products and reactants are equal.

Student Book
pages 38–39

Key points

- Using the right amount of raw materials prevents wastage and saves money.
- The cost of raw materials affects a wide range of consumer products.
- The cost of industrially made products depends on the costs of the reactants, overheads and waste.

2.10 The cost of a product

- If ammonia can be made more cheaply, then it costs less to produce ammonium nitrate (a fertiliser). This makes it cheaper for farmers to fertilise their land, for more crops to be grown, and for food to be made more cheaply. All consumer products are affected in this way.

Here are some factors that affect how much it costs to make products:

- Materials – not knowing how much reactant to use can waste money
- Wages – workers need to be paid
- Energy – the factory needs electricity and/or fuel
- Taxes – the government takes a share of all earnings
- Transportation – it costs money to deliver products and collect reactants
- Research – money is spent discovering new ways to develop products

⟼ **1** *Suggest three ways to reduce the cost of a product.*

1 Are these chemicals elements or compounds? CH_4, O_2, KI, Cl_2, HBr, H_2, $AgNO_3$, Br_2, Ca

2 If a hydrogen atom loses an electron, what charge will the resulting ion have?

3 Why are we able to separate the hydrocarbons in crude oil?

4 What are the three main chemical reactions involved in extracting iron from iron ore?

5 Why can't carbon be used to extract aluminium from its ore?

6 Name three gases extracted from the atmosphere, and give two uses for each of them.

7 How can a mine damage: **a** the surrounding landscape; **b** the atmosphere?

8 Suggest two ways of reducing the negative impact of opening a mine. (Higher tier: three ways)

9 What happens to the reactants during a chemical reaction?

10 Look at this chemical reaction and answer the questions below: $A + B \longrightarrow X + Y$

    a Name the products.

    b If 23 g of A and 32 g of B produce 12 g of X, how much Y is made?

11 Which one of these equations isn't balanced? For the unbalanced one, explain what is wrong and balance it.

    a $Mg + H_2SO_4 \longrightarrow MgSO_4 + H_2$ **b** $Mg + HCl \longrightarrow MgCl_2 + H_2$ [H]

12 Give four reasons why the price of food can increase.

Chapter checklist ✓✓✓

Tick when you have:

reviewed it after your lesson ✓ ☐ ☐

revised once – some questions right ✓ ✓ ☐

revised twice – all questions right ✓ ✓ ✓

Move on to another topic when you have all three ticks

Building blocks of new products	☐	☐	☐
Inside atoms	☐	☐	☐
Different types of particles	☐	☐	☐
Making products with materials from the Earth	☐	☐	☐
Extracting metals for construction	☐	☐	☐
Products from the atmosphere	☐	☐	☐
Exploiting the Earth's resources	☐	☐	☐
Changing materials to make new products	☐	☐	☐
Using equations	☐	☐	☐
The cost of a product	☐	☐	☐

**Student Book
pages 44–45**

Key points

- Classification sorts things that share similar features into groups.
- Organisms are classified into groups based on their physical characteristics.
- Organisms are classified to aid naming and identification.

AQA Examiner's tip

Knowledge of the specific characteristics that classify organisms into groups is **not** required for the exam. However, you should recognise that if one organism shares physical features with another, they can be classified into the same group.

3.1 Classification

- **Classification** means sorting things into groups based on similar features.

⯈ **1** *What is 'classification'?*

- Living organisms are classified into taxonomic groups. These range from **kingdom** (organisms share some characteristics) to **species** (organisms' characteristics are almost identical).

The main categories of living things include:

- Plants – organisms that make their own food by **photosynthesis**.
- Animals – organisms that cannot make their own food.
- Microbes – includes fungi and single-celled organisms.

⯈ **2** *What is the main characteristic of the plant kingdom?*

Scientists classify organisms to:

- Name and identify species
- Predict characteristics
- Find evolutionary links.

Key words: classification, species

**Student Book
pages 46–47**

Key points

- An evolutionary tree shows the evolutionary relationship between species.
- The size of predator populations depends upon the population of their prey.

AQA Examiner's tip

When answering questions about evolutionary trees, you must look at the drawings of the organism, and the time line. Organisms alive today are normally drawn at the top of the diagram and the common ancestor found at the bottom. If a species is not present in the most recent level of the diagram, it is extinct.

3.2 Using evolutionary and ecological relationships

- Fossil records have enabled scientists to produce **evolutionary trees**. These are branched diagrams that show how different species have evolved from a common ancestor. Evolutionary trees are produced by looking at similarities and differences in a species' physical characteristics and genetic makeup.

⯈ **1** *What does an evolutionary tree show?*

The elephant's evolutionary tree

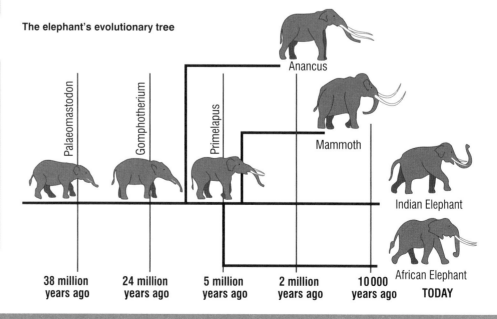

| 38 million years ago | 24 million years ago | 5 million years ago | 2 million years ago | 10 000 years ago | TODAY |

● Scientists also study how species depend on each other. These interactions are known as **ecological relationships**: for example, the interactions between **predators** and their **prey**.

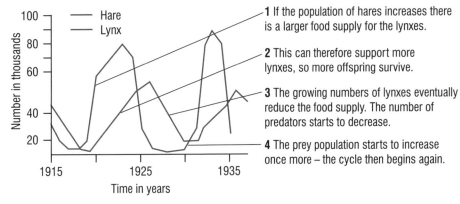

1 If the population of hares increases there is a larger food supply for the lynxes.

2 This can therefore support more lynxes, so more offspring survive.

3 The growing numbers of lynxes eventually reduce the food supply. The number of predators starts to decrease.

4 The prey population starts to increase once more – the cycle then begins again.

Changes in the populations of lynx and snowshoe hares, 1915–35

▓▶ **2** *What is meant by an 'ecological relationship'?*

Key words: evolutionary tree, ecological relationship, predator, prey

---

Student Book pages 48–49

Key points

● A habitat is a place where an organism lives.
● Plants need sunlight, water and nutrients to survive.
● Animals need food, water, mates and a suitable territory to survive.

Bump up your grade

There are lots of key words in this chapter. For example, community, ecosystem, habitat, environment, population and species. Try making your own ecological dictionary to learn their definitions.

3.3 Competition

● A **habitat** is the place where an organism lives. For example, the ocean, forest or pond. Each habitat has different environmental conditions. These include temperature, and amount of rainfall. The environmental conditions in most habitats vary throughout the day and throughout the year.

▓▶ **1** *What is a habitat?*

● An **ecosystem** is the name given to a habitat and all the organisms that live there. To survive, the plants and animals need materials from their surroundings. If materials are limited, organisms compete for these resources – only the best adapted species will survive.

▓▶ **2** *What is an ecosystem?*

● To survive, plants need sunlight and water (for photosynthesis) to produce food for growth. They also need nutrients. Animals need food and water (to grow), mates (to reproduce) and suitable territory.
● Many different plants and animals can be found in a pond. The different **populations** of species live together in a **community**. The variety and number of organisms in this community are determined by the materials available and the interactions between the organisms present.

▓▶ **3** *What is a community?*

Key words: habitat, ecosystem, population, community

Student Book
pages 50–51

Key points

- An adaptation is a specific characteristic that allows an organism to live in a particular habitat.
- Plants adapt to their surroundings through changes in surface area, water storage tissues and extensive root systems.
- Animals adapt to their surroundings using surface area, insulation, body fat and water storage.

AQA Examiner's tip

The surface area of an animal, compared with its volume, affects how quickly an animal loses heat energy. Do not assume that a large animal has a large surface area – this is not always the case.

3.4 Adaptations

- All organisms have characteristics that allow them to live successfully in their habitat. These characteristics are known as **adaptations**.

▐▐▶ **1** *What is an adaptation?*

A prairie crocus has specific adaptations to help them survive in the Arctic. The plants are small and grow close together to protect them from cold and strong winds. The stems, leaves and buds are covered in fine hairs that provide insulation. They have small leaves to reduce water loss.

Adaptation	Camel – desert	Polar bear – Arctic
Surface area to volume ratio	Large to maximise heat loss	Small to reduce heat loss
Fur	Thick on top of the body for shade, thin everywhere else to maximise heat loss	Thick layers of fat and fur – provides insulation; white – camouflage
Feet	Large and flat – stops them sinking into the sand	Large, with hairs on sole – avoids sinking into snow; hairs provide grip
Other adaptations	Do not sweat or urinate very often – reduces water loss from the body	Sharp claws and teeth – weapons to catch and eat prey

▐▐▶ **2** *What adaptations does a camel have to keep cool?*

- **Extremophiles** are microorganisms that can survive in extreme **environments**. These include volcanic vents, deep under the oceans, and highly acidic hot springs

Key words: adaptation, extremophile, environment

Student Book
pages 52–53

Key points

- Evolution is the theory that all organisms have evolved from a common ancestor.
- The fossil record provides evidence for evolution.
- Organisms evolve through natural selection.

3.5 Evolution

- All living organisms have gradually developed, from a common ancestor, over millions of years. This process is called **evolution**.

▐▐▶ **1** *What is meant by the term 'evolution'?*

- Fossils were formed when animal and plant remains were preserved in rock. The fossil record provides most of the **evidence** for evolution.
- Organisms evolve through the process of **natural selection**. They gradually change over time to become more adapted to their environment. It follows the steps below:

Organisms in a species show a wide range of **variation** (caused by genetic differences).

↓

The organisms with the characteristics that are most suited to the environment are most likely to survive and reproduce.

↓

Genes from successful organisms are passed to the offspring in the next generation. This process is then repeated over a number of generations.

▐▐▶ **2** *What is meant by 'natural selection'?*

**Student Book
pages 54–55**

- Peppered moths are a classic example of natural selection. Two varieties exist – a pale, light brown variety and a dark variety. The more camouflaged variety will survive and reproduce. Originally the pale moths were more widespread. However, after the Industrial Revolution dark moths became more common in towns and cities than pale moths.
- If a species is poorly adapted to its environment, it will not survive and will eventually become **extinct**. Extinction occurs naturally, but can be caused by human activity.

Key words: evolution, evidence, natural selection, extinct

Key points

- Light and temperature (day length), and gravity are some environmental factors that affect plant growth.
- Plants respond to their environment using auxins (plant hormones).
- Phototropism means growing towards the light. Gravitropism means growing towards gravity.

3.6 Plant growth

- Plants respond to their environment using **auxins**. These chemicals are hormones. Plants detect stimuli in their environment, and can respond by growth – **tropism**.

> **1 What is an auxin?**

- **Phototropism** means growing towards light. This means the plant can photosynthesise more. More food is produced for the plant, so it can grow faster.
- **Gravitropism** means growing towards a source of gravity. Growing deeper into the soil helps to provide anchorage. It normally takes the roots nearer to a source of water.

> **2 What is the difference between phototropism and gravitropism?**

- Auxins are made in cells near the tips of plant shoots or roots. They make these parts of the plant grow faster than others.

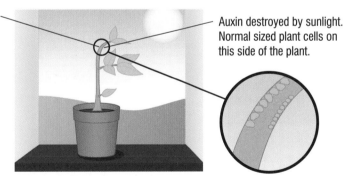

More auxin present on shaded side of plant. This causes cells to lengthen on this side of the plant.

Auxin destroyed by sunlight. Normal sized plant cells on this side of the plant.

Auxins cause plant shoots to grow towards the Sun

> **3 How do auxins work?**

Plant hormones that help crops to grow effectively include:

- Weedkillers can use growth hormones to selectively kill weeds, by making them grow too fast.
- Rooting powder (a growth hormone) is pasted onto plant cuttings to promote root growth.
- Ethene (a plant hormone) is released onto fruits so they ripen quicker.

Key words: auxin, tropism, phototropism, gravitropism

1 What is meant by the term habitat?

2 What is an extremophile?

3 What three things do plants need to survive?

4 Name three resources animals compete for.

5 What is gravitropism, and why is it important for a plant's survival?

6 What is the difference between an ecosystem and a community?

7 Describe and explain three polar bear adaptations, which allow it to live in the Arctic.

8 Why is it useful for scientists to classify organisms?

9 Describe how a plant uses auxins to grow towards the light.

10 Describe the steps which occur during natural selection.

Chapter checklist			✓	✓	✓
Tick when you have:			Classification	☐	☐	☐
reviewed it after your lesson	✓ ☐ ☐		Using evolutionary and ecological relationships	☐	☐	☐
revised once – some questions right	✓ ✓ ☐		Competition	☐	☐	☐
revised twice – all questions right	✓ ✓ ✓		Adaptations	☐	☐	☐
Move on to another topic when you have all three ticks			Evolution	☐	☐	☐
			Plant growth	☐	☐	☐

**Student Book
pages 58–59**

Key points

- Biomass is the name given to all the living organic matter present in an area.
- Producers make their own food by photosynthesis. Consumers have to eat food to gain energy.
- Food chains and webs show the flow of energy and biomass between organisms in an ecosystem.

4.1 Biomass and food chains

- **Biomass** is the mass of all living material found in an **ecosystem**. An ecosystem contains all the living matter (such as plants and animals) and non-living matter (such as rocks and water) in an area.
- Living organisms can be divided into two groups – **producers** and **consumers**.

▷ 1 What is biomass?

- Producers (plants and algae) make their own food. Energy from the Sun enters the **biosphere** as light energy. Producers absorb this and convert it into chemical energy, through the process of photosynthesis. They store it as organic compounds, such as carbohydrates, that can be converted into sugars, fats and proteins. These are used for growth, repair and as a source of energy.
- Photosynthesis can be summarised by the following equation:

$$\text{carbon dioxide} + \text{water} \xrightarrow{\text{light energy}} \text{glucose (sugar)} + \text{oxygen}$$

$$6CO_2 + 6H_2O \longrightarrow C_6H_{12}O_6 + 6O_2$$

- Consumers (animals) have to eat other organisms to gain energy. They gain energy from their food (biomass) through the process of **respiration**.
- Respiration can be summarised by the following equation:

$$\text{glucose} + \text{oxygen} \longrightarrow \text{carbon dioxide} + \text{water} \ (+\text{energy})$$

$$C_6H_{12}O_6 + 6O_2 \longrightarrow 6CO_2 + 6H_2O \ (+\text{energy})$$

▷ 2 What is the difference between a producer and a consumer?

- A food chain shows what organisms eat. The arrows in a food chain show the movement of energy (stored in food) from one organism to the next. Each step in the food chain is known as a **trophic level**.

Food chains always begin with a producer	A rabbit is a prey organism – it is eaten by another animal	A fox is a predator organism – it eats other animals

A simple food chain

- In most ecosystems, animals will eat more than one type of organism. This can be illustrated on a **food web** – a series of interlinked food chains.

Key words: biomass, producer, consumer, respiration, trophic level, food web

AQA Examiner's tip

Remember, plants respire as well as photosynthesise. They carry out respiration to release energy from nutrient stores in their body.

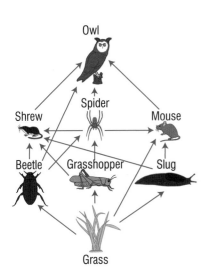

A food web

Student Book
pages 60–61

4.2 Energy transfer within food chains

- Only about 1% of the light energy coming from the Sun is transferred into chemical energy in a plant. Some is reflected back into the atmosphere, or used in photosynthesis reactions.
- The chemical energy gained is used for growth, increasing a plant's biomass, and providing food for consumers.

▐▐▶ **1** *What process converts light energy into chemical energy?*

- When one organism eats another, only around 10% of the energy is transferred. This results in less energy being available at each trophic level.

Energy is wasted because:

- Not all parts of a plant or animal may be eaten.
- Not all parts of the plant or animal can be digested.
- Energy released by respiration is used for movement and other body processes. It is eventually transferred to heat energy, and lost to the atmosphere.
- Energy is lost from the body in faeces and urine (waste products).

▐▐▶ **2** *Approximately how much energy is transferred from one trophic level to the next?*

Key points

- Approximately 10% of energy is transferred from one level of a food chain to the next.
- Not all light energy is converted into chemical energy by producers during photosynthesis – some is reflected or the wrong wavelength.
- Not all energy is passed to the next organism by consumers – some is 'lost' through respiration and waste.

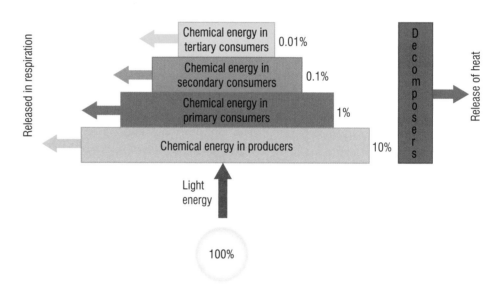

Energy flow through a typical food chain

AQA Examiner's tip

Remember that energy is never lost – it is transferred to a less useful form, such as heat, which can no longer be used for growth.

Maths skills

Calculating energy transfer

To calculate how much energy is transferred from one organism to the next, you can use the following equation:

$$\text{energy transferred} = \text{energy taken in} - \text{energy transferred in waste} - \text{energy transferred through respiration}$$

Or, expressed as a percentage:

$$\% \text{ energy transferred} = \frac{\text{energy transferred}}{\text{total energy taken in}} \times 100$$

4.3 Pyramids of numbers and biomass

Student Book pages 62–63

Key points

- Pyramids of numbers represent the number of organisms present at each trophic level.
- Pyramids of biomass represent the total amount of biomass present at each trophic level.

Sparrowhawk 0.1 kg
Blue tits 1 kg
Caterpillars 10 kg
Oak tree 100 kg

A pyramid of biomass – these are never inverted

Key words: pyramid of numbers, pyramid of biomass

- A **pyramid of numbers** shows the population at each level in a food chain. The width of each bar represents the number of organisms present. The diagram is usually pyramid-shaped, as an organism normally eats more than one organism from the trophic level below. As you move from one trophic level to the next, the size of organisms generally increases. However, there are fewer and fewer organisms at each level.

▶ **1** *What generally happens to the number and size of organisms as you move up a food chain?*

- Some pyramids of numbers are inverted, as these diagrams do not take into account the *size* of the organisms present. For example, a single tree can support a large number of living organisms.
- Population data can also be represented in a **pyramid of biomass**. The width of a bar in the diagram represents the biomass of organisms in the trophic level. This takes into account both the *number* and *size* of the organisms present.

▶ **2** *How is a pyramid of biomass different from a pyramid of numbers?*

 Maths skills

Calculating biomass
The total biomass contained in a trophic level can be calculated by:

total biomass = mass of one organism × number of organisms

4.4 Recycling of nutrients

Student Book pages 64–65

Key points

- Decomposers are microorganisms that break down dead organic material.
- Detritivores are small animals that speed up decomposition by shredding organic material into smaller pieces.
- Nutrients are constantly cycled through the ecosystem, so they can be used over and over again.

AQA *Examiner's tip*

Use a diagram, whenever possible, within an examination answer to add detail to your response.

- **Decomposers** are microorganisms (bacteria and fungi). They have **enzymes** that break down dead organic material and animal waste into soluble nutrients. These are absorbed into their bodies and used for growth and energy. If the microorganisms are eaten, the nutrients are passed on. Some nutrients are also released directly into the environment where they can be absorbed by plants.
- Decomposers work most efficiently in conditions that are warm, moist and oxygen-rich.

▶ **1** *What is a decomposer?*

- **Detritivores** are small animals, for example earthworms, woodlice and maggots that aid in the breakdown of organic material. They speed up decomposition by shredding the dead material into very small pieces, which makes it easier for decomposers to break down.

▶ **2** *What is a detritivore?*

- Plants obtain nutrients from the soil. These are then passed on to animals when the plant is eaten. When plants and animals die, decomposers release the nutrients trapped in them back into the soil, where they are absorbed by plants. This process is known as nutrient cycling.

Key words: decomposer, enzyme

4.5 The carbon cycle

Key points

- Carbon is constantly cycled throughout the biosphere.
- Carbon dioxide is removed from the atmosphere through photosynthesis.
- Carbon dioxide is released into the atmosphere through respiration and burning.

Bump up your grade

If you have to answer a question about the carbon cycle, use a drawing to show the relationships between photosynthesis, respiration, combustion and decomposition.

- The **carbon cycle** shows how carbon is cycled through the environment.
- Carbon dioxide is removed from the atmosphere during photosynthesis.
- Plants use light energy to convert carbon dioxide and water into glucose and oxygen.
- Glucose can then be used to make complex carbohydrates, fats and proteins that are needed for growth (turning the carbon into extra biomass).
- When plants are eaten, carbon is transferred to the animal where it can be used to produce fats and proteins.

▌▶ **1** *How is carbon removed from the environment?*

Carbon is released into the atmosphere in the following ways:
- **Respiration** – organisms respire to release energy from their food. This gives off carbon dioxide back into the atmosphere.
- **Decomposition** – decomposers release carbon dioxide when they break down organisms' remains.
- **Combustion** – when fossil fuels are burned, trapped carbon is released back into the atmosphere as carbon dioxide. Fossil fuels include coal, oil and natural gas.

▌▶ **2** *How is carbon released into the environment?*

Key words: carbon cycle

4.6 Human influence on the carbon cycle

Key points

- Carbon may be stored in the bodies of plants and animals, fossil fuels and limestone rock.
- Limestone forms from the shells and skeletons of marine animals.
- Deforestation and the burning of fossil fuels increase the levels of carbon dioxide in the atmosphere.

- Carbon is stored away from the atmosphere:
  - In the oceans.
  - Temporarily in plant and animal bodies, e.g. rain forests.
  - In fossil fuels. The carbon is released when the fuel is burned.
  - In rocks, such as limestone. Carbon may be stored for millions of years in a 'carbon sink'.
- Marine animals (such as coral) convert carbon dioxide dissolved in sea water into calcium carbonate to make shells and bones. When these organisms die, their shells are deposited on the sea bed. Eventually this sediment builds up, forming limestone. Through chemical weathering of the rock or thermal decomposition, carbon may be released back into the atmosphere as carbon dioxide.

▌▶ **1** *Which parts of marine animals might form limestone?*

Human activity has increased atmospheric carbon dioxide levels through:
- Burning fossil fuels – populations increase and higher living standards require more energy. More fuels have to be burned to generate this electricity.
- Deforestation – large areas of forest have been cleared for roads, buildings and agriculture. This results in fewer trees removing carbon dioxide from the atmosphere by photosynthesis. Burning the trees further increases the levels of atmospheric carbon dioxide.

1 What is the difference between a producer and a consumer?

2 Name three stores of carbon.

3 Using the food web on page 19, answer the following questions:

    a What does a beetle eat?

    b What does a spider eat?

4 What is the difference between a decomposer and a detritivore?

5 **a** Name a situation where a pyramid of numbers may be inverted.

    b Why does the situation in part a occur?

6 Name three reasons why energy is not transferred from one organism to the next in a food chain.

7 What is meant by the term 'nutrient cycling'?

8 Why does an increasing use of energy lead to an increase in atmospheric carbon dioxide?

9 A rabbit has taken in 120 kJ of energy from its food. Of this 75 kJ are lost from the food chain through waste, and 35 kJ are lost through respiration. If a fox eats a rabbit, how much energy from the food is transferred from the fox to the rabbit? State your answer as:

    a an amount in kilojoules

    b a percentage.

Chapter checklist ✓ ✓ ✓

Tick when you have:				Biomass and food chains	☐	☐	☐
reviewed it after your lesson	☑	☐	☐	Energy transfer within food chains	☐	☐	☐
revised once – some questions right	☑	☑	☐	Pyramids of numbers and biomass	☐	☐	☐
revised twice – all questions right	☑	☑	☑	Recycling of nutrients	☐	☐	☐
Move on to another topic when you have all three ticks				The carbon cycle	☐	☐	☐
				Human influence on the carbon cycle	☐	☐	☐

Unit 1

1 Astronomers have noticed that the frequency of infrared waves coming from the stars is lower than expected.

  a Explain why the frequency is lower than expected. *(2 marks)*

  b What does this evidence lead the astronomers to believe is happening to the universe? *(1 mark)*

2 The Earth's crust is cracked into large pieces called plates, which move about.

In this question you will be assessed on using good English, organising information clearly and using specialist terms where appropriate.

  a Explain why earthquakes occur. *(6 marks)*

  b How did the early volcanoes change the Earth's atmosphere? *(2 marks)*

3 The level of carbon dioxide in the atmosphere has changed a lot over the history of the Earth.

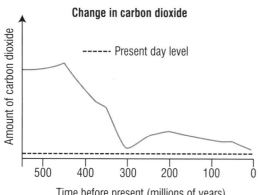

Change in carbon dioxide

------ Present day level

Amount of carbon dioxide

Time before present (millions of years)

500 400 300 200 100 0

  a Suggest why the level of carbon dioxide decreased around 450 million years ago. *(1 mark)*

  b Carbon dioxide is a greenhouse gas. Describe the need for some greenhouse gases on Earth. *(2 marks)*

4 The masses of the organisms in a food chain are given in the table.

Organism	Mass in grams
Otter	1000
Small fish	300
Water weeds	250
Insect larvae	10

  a Use the pyramid of biomass to calculate how many of each organism there are at each stage in the food chain. *(2 marks)*

Otter 1000 g/m²
Small fish 1500 g/m²
Insect larvae 2000 g/m²
Water weeds 5000 g/m²

  b Describe how to measure the biomass of the water weeds. *(2 marks)*

  c Give **three** ways in which energy is lost between the small fish and the otter. *(3 marks)*

AQA **Examiner's tip**

Make sure you think through your answer to question 2a clearly first so that you write it in a logical order. Put all the relevant specialist terms in, as the examiner will have a list of these to check your answer against.

5 Plants in the desert have very thin leaves so that they do not lose much water from them.

  a i Why is this adaptation of benefit to the plant? *(1 mark)*

     ii Explain how these plants have adapted over generations. *(3 marks)*

  b Explain what other adaptation plants living in the desert would need. *(2 marks)*

6 There are two types of peppered moth in Great Britain. One is pale and speckled and the other is dark. Both types are nocturnal and rest during the day on tree trunks. Before 1850 the dark variety was rare. During the late 1800s and early 1900s, pollution from heavy industries killed lichen that grows on the bark of trees and soot from factories was released.

  a Suggest and explain what happened to the population of pale speckled peppered moth after 1850. *(3 marks)*

In 1956 the Clean Air Act was put in place to reduce the amount of smoke pollution.

  b Suggest and explain which type of moth has the higher population now. *(5 marks)*

7 The air contains gases that, once separated from each other, are very useful.

  a Describe how these gases can be separated before use. **[H]** *(4 marks)*

  b Give **one** reason why the production of nitrogen is useful for farming. **[H]** *(1 mark)*

8 Iron ore (Fe_2O_3) needs to be chemically processed before the iron can be used.

  a Describe the process used to extract iron from iron oxide using coke. *(4 marks)*

  b Give a balanced symbol equation of the extraction of iron from its ore using carbon monoxide. **[H]** *(3 marks)*

AQA Examiner's tip

For the type of question shown in question 6, you need to make sure you have read and understood all of the information given in the question so that you use it in your answer.

Student Book
pages 76–77

5.1 Nerves

Key points

- Nerves transmit electrical impulses, so that the body can react to changes in your environment.
- During a controlled nervous reaction, information is sent to the brain to be processed.
- Reflex responses occur without thinking, so are much quicker than controlled responses.

- Your body responds to an external change (a **stimulus**) by sending electrical impulses around your body, using your nervous system.
- **Receptors** are groups of cells found in your sense organs (such as the eye and ear) that detect the stimulus. They include cells that detect light, sound, smell, taste, touch, and heat. Most information detected by your body is sent along **neurons** (which are bundled together in nerves) to the brain. The brain processes the information received.
- The brain then sends an impulse to the **effectors** (muscles or glands), causing a response.

➡ **1** *What are receptor cells?*

➡ **2** *Name three stimuli that the body responds to.*

There are three types of neuron (nerve cell):

- **Sensory neurons** that carry electrical impulses from receptor cells to the **central nervous system (CNS)**. The CNS is made up of the brain and spinal cord.
- **Relay neurons** (found in the CNS) that carry electrical impulses from sensory neurons to motor neurons.
- **Motor neurons** that carry electrical impulses from the CNS to effectors.

The diagram below shows the steps involved in a controlled nervous reaction to a stimulus:

Stimulus ⟶ Receptor cells ⟶ Sensory neuron ⟶ Spinal cord ⟶ Brain ⟶ Spinal cord ⟶ Motor neuron ⟶ Effector ⟶ Response

- Reflex actions are automatic reactions which occur without thinking. They do not involve the brain. Therefore the body can react more quickly. Reflex actions are used when a person is in danger.

The diagram below shows the steps involved in a reflex action:

Stimulus → Receptor cells → Sensory neuron → Spinal cord → Motor neuron → Effector → Response

AQA *Examiner's tip*

Make sure that you spend time learning the order of a nervous response, and the special vocabulary used. Don't forget that a reflex action removes the brain from the cycle – fewer steps in the sequence means a faster response.

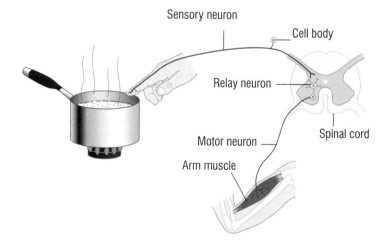

Reflex actions only take around 0.2 of a second

Key words: stimulus, receptor, neuron, effector, central nervous system (CNS)

➡ **3** *What is the difference between a controlled response and a reflex action?*

Key points

- Sound travels as longitudinal waves through solids, liquids and gases.
- Humans can hear sounds between 20 Hz and 20 000 Hz.
- Prolonged exposure to loud sounds can cause permanent damage to hearing.

5.2 Hearing

- Sound energy travels as a wave. For example, when a drum is struck it vibrates, causing the air particles next to the drum to vibrate. These **vibrations** pass energy onto neighbouring particles. This eventually causes your ear drum to vibrate, and so you hear the sound. As the vibration passes, the air particles are squashed together. This is called a **compression**. In other places air particles become spread out – a **rarefaction**.
- Sound waves are **longitudinal waves**. The particles vibrate in the same direction that the energy is travelling in. Sound travels quickly through solids, and slowly through gases.

▶ **1** *What type of wave is a sound wave?*

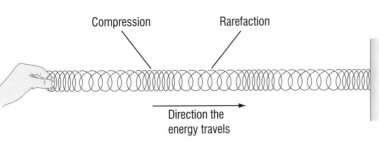

A longitudinal wave travelling along a slinky spring

- The speed at which something vibrates is called its **frequency**. One vibration per second is a frequency of one **hertz (Hz)**. The human hearing range is between around 20 Hz and 20 000 Hz.

▶ **2** *What is the human range of hearing?*

- Noise levels are measured in decibels (dB). Hearing loss can be caused by any sound above 85 dB. The damage caused depends on the loudness of the sound, and the length of exposure.

- Regularly listening to an MP3 player at high volume can damage your hearing. Initial damage is often reversible, but prolonged exposure can cause tinnitus (permanent ringing in the ears) and an inability to hear certain frequencies.

Key words: vibration, compression, rarefaction, longitudinal wave, frequency, hertz

---

Key points

- Hormones are chemical messengers that travel around your body in the blood.
- Hormones control body processes that need constant adjustment, to maintain a constant internal environment.
- Homeostasis is the maintenance of a constant internal environment.

5.3 Hormones

- **Hormones** are chemical messengers that travel around your body. They help to maintain a constant internal environment (**homeostasis)**, which is essential for the body to function normally.
- Hormones are made in **glands** and secreted into the blood. Hormones cause a response in specific cells found in **target organs**.
- Hormones and nerves carry out similar roles, but act in different ways. Nerves have an immediate effect in a very precise area, but the response is very short acting. In comparison, hormones cause a slower response over a larger area, but their effects are longer lasting.

▶ **1** *What are hormones and where are they made?*

The body maintains a steady state through **negative feedback**. This means that any changes which affect the body are reversed, and returned to normal. For example: receptor detects change in body temperature ⟶ brain receives information⟶ effector produces a response to return body temperature to normal.

Higher

Key words: hormone, homeostasis, gland, target organ, negative feedback

AQA Examiner's tip

When trying to explain a system such as negative feedback, it is a good idea to include a flow diagram in your answer.

Student Book
pages 82–83

5.4 Diabetes and controlling blood sugar levels

- Your body needs glucose (sugar) for energy. However, too much glucose in the blood can cause serious health problems.
- The hormones **insulin** and **glucagon** are responsible for maintaining a constant blood glucose level.

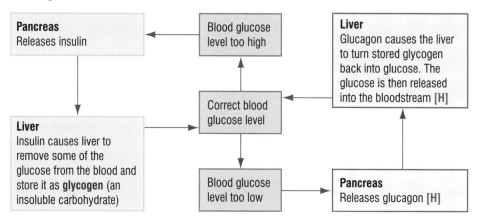

The diagram shows how insulin and glucagon work together to ensure a constant blood glucose level

1 *Which hormones are responsible for controlling blood glucose levels?*

People who cannot control their blood glucose levels suffer from **diabetes**. There are two main types:

- **Type 1** – sufferers do not produce insulin, so they have regular insulin injections and maintain a healthy diet and regular exercise.
- **Type 2** – sufferers do not produce enough insulin, or produce poor quality insulin. Sufferers avoid eating large quantities of carbohydrate-rich foods, and exercise after eating to use up excess glucose.

2 *What condition may a person suffer from if they can't control their blood glucose levels?*

Student Book
pages 84–85

5.5 Controlling body temperature

- The **thermoregulatory centre** in the brain is responsible for maintaining body temperature at 37 °C. A couple of degrees difference stops the body working efficiently. For example, if a person is too hot, they may suffer from fits and dehydration. If they are too cold, body movements will slow, eventually causing a coma and death.

1 *What is the normal body temperature?*

- When a change in body temperature is detected, the brain causes different parts of the body to respond. These responses should return the body back to its normal temperature.

What happens when you get too hot?

- Hairs on your skin lie flat.
- Sweat glands produce sweat. Sweat is mainly water, but it also contains salt and urea (a waste material). The water in sweat absorbs heat energy from your body to evaporate. As heat energy is lost from your body, your temperature falls so you feel cooler.
- Blood vessels supplying **capillaries** near the surface of your skin **dilate** (widen). This increases blood flow through the capillaries, increasing heat loss by radiation.

▐▐▐➡ **2** *What is sweat made of?*

What happens when you get too cold?

- Hairs on your skin stand on end, trapping a layer of air close to the skin, preventing heat loss.
- Sweat glands do not produce sweat.
- Blood vessels supplying capillaries near the surface of your skin narrow (constrict). This reduces blood flow through the capillaries, reducing heat loss.
- Shivering requires extra energy, so your cells respire more, producing extra heat.

▐▐▐➡ **3** *Why do the hairs on your arm stand up when you are cold?*

Key words: thermoregulatory centre, capillary, dilate

AQA Examiner's tip

If answering a question about maintaining body temperature in an exam, use your own skin as a reference. Ask yourself – if you are feeling hot or cold, how does your skin's appearance change?

Student Book
pages 86–87

Key points

- Chemical reactions control everything that happens in your body.
- Acids and bases are involved in many of your body's chemical reactions.
- Some acids and bases can damage skin, teeth and internal organs.

Bump up your grade

To improve your grade, make sure you can identify the hazard **and** explain what it means, describe how to reduce the risk and give examples.

5.6 Body chemistry

- Many chemical reactions go on in our bodies all the time. They support the seven life processes of movement, respiration, sensitivity, growth, reproduction, excretion and nutrition.
- Proteins, called **enzymes**, control most of these chemical reactions.
- Acids and bases are involved in many of your body's chemical reactions. For example, the stomach makes hydrochloric acid and your liver makes weakly alkaline bile. Bile neutralises stomach acid and helps break down fats in food.

▐▐▐➡ **1** *Name an acid and a base made in our bodies.*

- Acids and bases can be dangerous. They can cause painful chemical burns to skin or eyes. Acids can cause indigestion and heartburn, nausea and tooth decay.
- Strong acids and bases, for example, cleaning products, have warning signs on them called **hazard symbols**. Always take safety precautions when working with these (wear goggles, gloves or protective clothing).

Hazard symbols found on acids and bases

Key words: hazard symbol

Student Book
pages 88–89

5.7 Acids and bases

- Acids are chemicals that split up to release **hydrogen ions** (H^+) when dissolved in water. These ions take part in many chemical reactions. In your stomach these reactions involve breaking down proteins and killing bacteria.
- We can show hydrochloric acid splitting up like this:

$$HCl \longrightarrow H^+ + Cl^-$$
hydrochloric acid $\longrightarrow$ hydrogen ion + chloride ion

▶ 1 *What type of charge does a hydrogen ion have?*

- Bases are chemicals that react with hydrogen ions, usually forming a salt. This reaction is called **neutralisation**. An **alkali** is a type of base that dissolves in water.
- The pH scale measures how acidic or basic a solution is. Solutions at pH 7 are neutral, neither acidic nor basic. Chemicals, called indicators, can be used to distinguish acids and alkalis. They change colour depending on their pH. Universal indicator is red/orange/yellow in acidic solutions, blue/purple in alkaline solutions and green in neutral solutions. Matching the colour using the pH scale gives a pH value. Acids have a pH below 7 and alkalis have a pH above 7.

Key words: hydrogen ions, neutralisation, alkali

Key points

- Your stomach produces hydrochloric acid. This kills bacteria and helps digest food.
- Acids all release hydrogen ions (H^+) in solution.

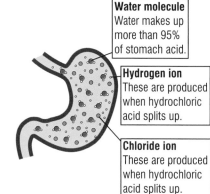

Water molecule
Water makes up more than 95% of stomach acid.

Hydrogen ion
These are produced when hydrochloric acid splits up.

Chloride ion
These are produced when hydrochloric acid splits up.

Hydrochloric acid in your stomach

Student Book
pages 90–91

5.8 Reacting acids with alkalis and bases

- Weak alkalis can be useful medicines. They neutralise excess stomach acid, relieving indigestion and heartburn.
- Like acids, alkalis split up when dissolved in water. They release **hydroxide ions** (OH^-). Sodium hydroxide is an alkali that splits up like this:

$$NaOH \longrightarrow Na^+ + OH^-$$
sodium hydroxide $\longrightarrow$ sodium ion + hydroxide ion

- We use alkalis like aluminium hydroxide $Al(OH)_3$ and magnesium hydroxide $Mg(OH)_2$ in **antacids** (weak bases) to treat excess stomach acid.

Hydroxide ions (in alkalis) react with hydrogen ions (in acids) to make water:

$$H^+(aq) + OH^-(aq) \longrightarrow H_2O(l)$$

The other ions from the acid and base react together to make a chemical called a **salt**.

▶ 1 *Use ideas about ions to explain why water molecules have no overall charge.*

To show how hydrochloric acid is neutralised by sodium hydroxide, we would write:

$$HCl + NaOH \longrightarrow H_2O + NaCl$$
hydrochloric acid + sodium hydroxide $\longrightarrow$ water + sodium chloride

If acid is reacted with a carbonate, then carbon dioxide is also produced:

$$2HCl + CaCO_3 \longrightarrow H_2O + CaCl_2 + CO_2$$
hydrochloric acid + calcium carbonate $\longrightarrow$ water + calcium chloride + carbon dioxide

Key points

- Hydroxide ions make solutions alkaline.
- Bases neutralise acids by reacting with the hydrogen ions they produce. They make a salt and water.

AQA Examiner's tip

Make sure you remember to put a + or a – to show the charge on an ion. Don't just write 'H ions', write 'H^+ ions'.

Key words: hydroxide ion, antacid

1 What is meant by 'homeostasis'?

2 What are the three main types of neuron?

3 Name three changes which take place in the skin when a person is too hot.

4 Name two differences between a controlled reaction and a reflex action.

5 Describe how you hear a sound when cymbals are crashed together.

6 **a** How does insulin control blood sugar levels?

    b How does glucagon control blood sugar levels? [H]

7 What is meant by the term 'negative feedback'? [H]

8 Describe three ways acids can harm your body.

9 Explain why toothpaste is a weak alkali.

10 What acid does your stomach produce, and why?

11 What are antacids?

12 Excess stomach acid (HCl) can be neutralised by magnesium hydroxide ($Mg(OH)_2$).

    a Write a word equation to describe this reaction.

    b Write a balanced symbol equation to describe this reaction. [H]

Chapter checklist

Tick when you have:

reviewed it after your lesson	✓	☐ ☐
revised once – some questions right	✓ ✓	☐
revised twice – all questions right	✓ ✓ ✓	

Move on to another topic when you have all three ticks

	✓ ✓ ✓
Nerves	☐ ☐ ☐
Hearing	☐ ☐ ☐
Hormones	☐ ☐ ☐
Diabetes and controlling blood sugar levels	☐ ☐ ☐
Controlling body temperature	☐ ☐ ☐
Body chemistry	☐ ☐ ☐
Acids and bases	☐ ☐ ☐
Reacting acids with alkalis and bases	☐ ☐ ☐

6.1 Animal cells

Student Book
pages 94–95

Key points

- Animal cells contain a nucleus, cell membrane, cytoplasm and mitochondria.
- Chromosomes are strands of DNA. On each chromosome the DNA is grouped into sections called genes that code for a specific characteristic.
- Parents' characteristics are passed on to their offspring through genes.

- Your body is made up of millions of **cells**. All animal cells have four main features:

A: **Cytoplasm** – this is a 'jelly-like' substance, where most chemical reactions occur.

B: **Cell membrane** – this controls what comes in and out of the cell.

C: **Nucleus** – this contains the information that determines a cell's appearance and function as well as the information needed to make new cells.

D: **Mitochondrion** (plural mitochondria) – this is where respiration occurs, producing energy.

You can see a typical animal cell in the diagram below.

�§▶ **1** *What does the nucleus do?*

- **Chromosomes** are long strands of **DNA** inside the nucleus. Each chromosome is divided into sections of DNA. These 'coding' sections are called **genes**. They contain the information needed to produce a characteristic, e.g. eye colour. One chromosome can contain thousands of genes.

▶ **2** *What is a gene?*

- We inherit characteristics from our parents through genes. During fertilisation, genes from the mother (carried in the egg cell) combine with genes from the father (carried in the sperm cell).

Key words: chromosome, DNA, gene

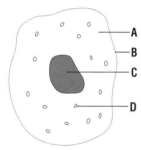

A typical animal cell

6.2 Variation

Student Book
pages 96–97

Key points

- Variation is the name given to the differences that exist between organisms of the same species.
- Variation occurs as a result of genetic factors, environmental factors, or a combination of both.

- Differences within a species are called **variation**. For example, people vary in many ways including height, build, hair colour and intelligence.

▶ **1** *What is variation?*

There are two factors that cause variation:
- The characteristics you **inherit** from your parents – genetic variation.
- The environment in which you live (environmental variation). Factors in a person's environment include where they live and their education.

Most characteristics are affected by both environmental and genetic variation, such as:
- Hair colour: generally children have a similar hair colour to one of their parents – genetic variation. However, people may dye their hair – environmental variation.
- Height: this is mostly determined by your genes. If your parents are tall, you are likely to be tall. However, your rate of growth can be affected as a result of a poor diet.

▶ **2** *Name two characteristics that are influenced by environmental variation.*

Key words: variation, inherited

**Student Book
pages 98–99**

6.3 Dominant and recessive alleles

● Different forms of the same gene are called **alleles**. Some genes will always be expressed (that is, their characteristic will appear in the individual). These are called **dominant** alleles. **Recessive** alleles will only be expressed if *both* the genes in a pair are recessive.

Key points

- Alleles are different forms of the same gene.
- Dominant alleles are always expressed if they are present in the nucleus. Recessive alleles are only expressed if a person has two copies of this allele.
- Punnett square diagrams are used to work out the chance of a characteristic being inherited.

▐▐▐➤ **1** *What is the difference between a dominant and recessive allele?*

● Punnett squares are used to explain how you inherit eye colour.
● A person's genes can be represented using letters. In the example below, **B** = brown eyes allele (dominant), and **b** = blue eyes allele (recessive).

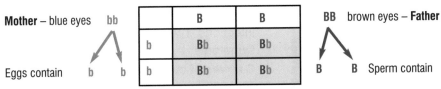

Studying the inheritance of one gene is called monohybrid inheritance

● All children in this example will have the genes **Bb**. This means that they will all have brown eyes, as **B** is the dominant allele.

So what would happen if both your parents had brown eyes, but they also carried the recessive gene for blue eyes?

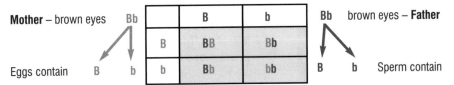

During fertilisation, any of the combinations shown in the yellow area of the table could be produced

● Children would be born in the ratio of 1 **BB** : 2 **Bb** : 1 **bb**. The chances are, for every three children born with brown eyes, there will be one child born with blue eyes – a 25% chance that the child will have blue eyes.

▐▐▐➤ **2** *What is the missing number in this statement about probability? A 25% chance is the same as a one in _____ chance.*

Key words: allele, dominant, recessive

Student Book
pages 100–103

6.4–6.5 Genetically inherited disorders

- Genetically inherited disorders are passed down from parents to their children in their genes. Examples include cystic fibrosis and sickle-cell anaemia.
- If the risk of having a child with a genetically inherited disorder is high, doctors may genetically screen a fetus. The test detects the presence of the genes responsible for these disorders. If the gene is detected, a couple may decide to abort the fetus.

> **1** *Why do doctors genetically screen some fetuses?*

- Cystic fibrosis is caused by a 'faulty' gene – a recessive allele. A person will only suffer from the disease if both their copies of the gene are faulty. Sufferers produce thick sticky mucus, which causes breathing difficulties and chest infections. There is no cure, but physiotherapy and antibiotics help manage symptoms.
- **Carriers** of cystic fibrosis have one copy of the faulty gene and one normal gene. They are healthy, but can pass the disorder on to their children if their partner also has a copy of the faulty gene.

> **2** *Why do cystic fibrosis carriers show no symptoms of the disease?*

- Sickle-cell anaemia is caused by a recessive allele. The allele causes red blood cells to become sickle-shaped instead of round. They cannot transport oxygen properly, resulting in anaemia. The condition is managed by pain relief, antibiotics and blood transfusions.
- Polydactyly is caused by a dominant allele. A person only needs one copy of the allele to suffer from this disease. People cannot be carriers, if you have the allele you will have the disease. Sufferers of polydactyly are born with extra digits on their hands or feet. They can be removed by surgery.

> **3** *What are the symptoms of polydactyly?*

- Haemophilia only affects males. It is caused by a recessive allele carried on the X chromosome (one of the sex chromosomes).
- Haemophilia stops the blood clotting; sufferers often experience internal bleeding. This also means that, without treatment, a cut could result in death. Regular injections of Factor 8 (a blood clotting factor) are given to control the condition.

> **4** *Which chromosome carries the haemophilia-causing allele?*

Most genetic disorders have no cure. Two possible treatments may include:
- **Gene therapy** – this has the potential to replace 'faulty' alleles with normal, healthy alleles in the tissues where the disease causes damage.
- **Stem cell** transplants – stem cells can grow into any cell type in the body. They can be removed from an embryo or umbilical cord. It may be possible to transplant healthy stem cells into a person who suffers from a genetic disorder.

> **5** *What are stem cells?*

Key words: carrier, gene therapy, stem cell

Key points

- Genetically inherited disorders are passed from parents to their children in their genes.
- Cystic fibrosis sufferers produce excessive levels of mucus, causing chest infections and difficulty absorbing food.
- Sickle-cell anaemia sufferers have sickle-shaped red blood cells that stop them carrying oxygen properly.
- Polydactyly is caused by a dominant allele. It causes children to be born with extra digits.
- Haemophilia prevents blood clotting properly. It is caused by a recessive allele carried on a sex chromosome.
- Gene therapy and stem cell transplants may be able to cure genetically inherited disorders in the future.

Bump up your grade

It may be helpful to use two different colour pens when drawing genetic crosses. One colour for the mother's genes and one colour for the father's genes. Then you can check that you have one gene from each parent.

AQA Examiner's tip

You are **not** expected to know whether a disorder is caused by a dominant or recessive allele. However, you must be confident that you understand the difference between the two types of allele.

1 A student is observing an animal cell through a microscope:

   a What is the outside of the cell called?

   b Where in the cell do chemical reactions take place?

   c Which part of the cell contains DNA?

2 What is meant by the term 'variation'?

3 What is the difference between a dominant allele and a recessive allele?

4 Place this sequence of structures in order of size (starting from the smallest):
cell, gene, nucleus, chromosome, DNA

5 Sort the following characteristics into those which show genetic variation,
environmental variation or both: blood group, height, intelligence, weight, eye
colour, sporting ability

6 Sickle cell anaemia is a disease caused by a recessive allele. This allele is
given the symbol 's'. The healthy allele is given the symbol 'S'.

   a Which alleles would a sufferer have?

   b Which alleles would a carrier have?

   c Many carriers do not know that they have the recessive allele. Why is this?

7 Polydactyly is a genetically inherited disorder caused by a dominant allele (P).
The healthy allele is recessive (p).

   a A man (Pp) and a woman (pp) wish to have a child. Use a Punnett square to
show the possible genetic make-up of their offspring.

   b What percentage of their offspring is likely to suffer from polydactyly?

Chapter checklist	✔ ✔ ✔
Tick when you have:	Animal cells ☐ ☐ ☐
reviewed it after your lesson ✔ ☐ ☐	Variation ☐ ☐ ☐
revised once – some questions right ✔ ✔ ☐	Dominant and recessive
revised twice – all questions right ✔ ✔ ✔	alleles ☐ ☐ ☐
Move on to another topic when you have all three ticks	Genetically inherited disorders ☐ ☐ ☐

Student Book
pages 108–109

7.1 Limestone as a building material

- **Limestone** is the starting point for many useful materials and is also a building material in its own right. Limestone contains mainly **calcium carbonate** ($CaCO_3$) and was formed from the shells of ancient sea creatures.

- Limestone is blasted from the ground in quarries. Once extracted, limestone can be built with directly or used to make other materials. Materials made from limestone include: **cement**, **mortar**, **concrete** and **glass**.

Key points

- Limestone (containing calcium carbonate, $CaCO_3$) is blasted from the ground in quarries.

- Limestone is used to make buildings, concrete, cement, mortar and glass.

- Cement is a starting point for making mortar and concrete.

- Mortar is the material that binds bricks together in a building.

- Concrete is used for making the structures of buildings.

- Glass is made from sand, with limestone and sodium carbonate added.

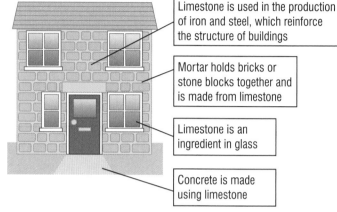

Limestone is used in the production of iron and steel, which reinforce the structure of buildings

Mortar holds bricks or stone blocks together and is made from limestone

Limestone is an ingredient in glass

Concrete is made using limestone

Most homes couldn't be built without limestone

▐▐▐▶ **1** *Which parts of homes are made using limestone?*

Key words: limestone, calcium carbonate

Student Book
pages 110–111

7.2 Limestone as a starting point

- **Quicklime** (calcium oxide) and **slaked lime** (calcium hydroxide) are both produced from limestone. They're used for making other building materials like cement.

- To make quicklime, limestone is crushed and then roasted in a kiln. This releases carbon dioxide.

Key points

- Quicklime is made by heating limestone to high temperatures.

- Slaked lime is made by adding water to quicklime. This reaction releases energy.

- The chemical name for quicklime is calcium oxide. Its chemical formula is CaO.

- The chemical name for slaked lime is calcium hydroxide. Its chemical formula is $Ca(OH)_2$.

limestone (calcium carbonate) $\xrightarrow{\text{heat}}$ quicklime (calcium oxide) + carbon dioxide

$$CaCO_3 \longrightarrow CaO + CO_2$$

▐▐▐▶ **1** *What is the waste gas released when quicklime is produced?*

- Slaked lime can be used to change the pH in soil and lakes affected by acid rain. I is made by carefully adding water to quicklime. A lot of heat energy is released in the process.

quicklime (calcium oxide) + water $\longrightarrow$ slaked lime (calcium hydroxide)

$$CaO + H_2O \longrightarrow Ca(OH)_2$$

Key words: quicklime, slaked lime

Student Book
pages 112–113

7.3 Products of limestone at work

- Limestone can be used as a construction material by simply stacking up big blocks of it. It is also used to make **cement**, which is then used to make **mortar** and **concrete**.

- Cement is made by heating powdered limestone with clay and gypsum in a rotating kiln. It is added to water and other materials. When it sets, it forms a hard, strong solid.

- Cement is used to make mortar and concrete. Mortar is made by adding sand and water to cement. It is strong, cheap and used to hold bricks together in buildings. Concrete is similar to mortar, but small stones are also added. Concrete is very strong and is often used to build whole buildings.

> **1** *What are the similarities and differences between mortar and concrete?*

- **Glass** is made by heating **sand**, limestone and **sodium carbonate** to about 1500°C. The molten glass can be made into different shapes before it cools down and sets. Glass is strong and hard, but not flexible (it is brittle).

Key words: cement, mortar, concrete

Key points

- Limestone is heated with clay and gypsum to make cement.
- Sand can be added to cement and mixed with water to make mortar. Concrete is made by mixing sand, cement and stones together.
- Sand, limestone and calcium carbonate are heated together to make glass.

AQA Examiner's tip

Don't confuse concrete with cement. Remember: cement is the material concrete is made from.

Student Book
pages 114–115

7.4 Metals for construction

- Metals are used extensively in our homes. Some examples are: electrical wires, reinforcement to add strength, pipes to carry water and gas, flashing on roofs, frames for doors and windows.

- Most metals share similar properties that make them ideal for use in the building industry.

Property	Meaning
Hard	Metals are difficult to scratch
Strong	Metals don't break easily
Malleable	Metals can be beaten into different shapes
Ductile	Metals can be pulled out into wires
High melting point	Most metals don't melt easily
Good conductors	Heat and electricity can travel through them easily
High density	They are heavy for their size because their atoms are packed closely together

- Metals have many uses in constructing homes. You don't always see them because they are often covered up.

- Different metals have different advantages for particular uses. For example, copper is a very good thermal and electrical conductor, it is ductile, has a high melting point, and is not very reactive. This makes it useful for making pipes and wires. Aluminium is lightweight and corrosion-resistant. This makes it useful for making window frames or pans.

> **1** *What properties make copper a good choice for wiring?*

Key points

- Metals all share similar properties. They are malleable, ductile, strong, hard, have high melting points and are good conductors.
- The properties of metals make them useful for many construction tasks.

Bump up your grade

Being able to state examples of metals used for building homes, like steel for reinforcement and copper for pipes, is a basic skill. For higher marks you need to explain **how** the metal's properties make it useful.

Student Book
pages 116–117

7.5 Polymers in the home

- **Polymers** are materials such as plastics. They are produced by making lots of small molecules (**monomers**) join together in a reaction to form long tangled chains.

The properties of polymers make them very useful in the home:

- low density (lightweight)
- flexible (usually)
- can be moulded into different shapes
- low melting point
- waterproof
- chemically unreactive
- insulators of heat and electricity.

The chemical reaction shown opposite is called **polymerisation**:

$$nC_2H_4 \longrightarrow (C_2H_4)_n \quad n = \text{number of monomer molecules}$$

| ethene | polyethene | (usually thousands) |
| (monomer) | (polymer) | |

1 *Write a word equation to show a polymer being made from propene.*

Some common polymers, their properties and their uses:

Name of polymer	Polyethene	Polystyrene	Polypropene	PVC (polyvinylchloride)
Properties	Cheap, light and flexible	Soft, lightweight insulator	Hard and rigid	Cheap and flexible
Uses	Bags, wrappings, toys	Insulation and packaging	Car bumpers, high pressure pipes	Water pipes, shower curtains

Key words: polymer, monomer, polymerisation

Key points

- Polymers are made from the products of crude oil.
- Polymers are easy to mould and are excellent insulators.
- Polymers are also waterproof and lightweight (have a low density).

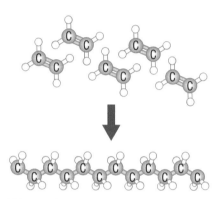

This reaction shows how part of a polyethene chain is made

AQA Examiner's tip

The name of most polymers is the same as the monomer it is made from, but with the word 'poly' in front of it. For example, polytetrafluoroethene is made from tetrafluoroethene.

Student Book
pages 118–119

7.6 Ceramics and composites in the home

- **Ceramics** are materials like clay or china and have been around for thousands of years. Tiles, sinks, toilets, plates and mugs are often made of ceramics.

The properties of ceramics are:

- good insulators of heat and electricity
- strong
- very hard
- unreactive
- **brittle**
- very high melting points.

1 *Why are ceramics used for heat shielding on spacecraft?*

Key points

- Ceramics are hard but brittle. They are good insulators. They are resistant to chemicals and have a high melting point.
- Composites are mixtures of more than one material. They often combine the best properties of each material.

● **Composites** are mixtures of materials. They combine different materials' best properties.

Composite name	Materials it is made of	Properties	Uses
Glass reinforced plastic (fibreglass)	Glass and resin	Light, strong, flexible	Water slides, speedboats
Reinforced concrete	Concrete, steel bars	Very strong, slightly flexible	Buildings
Reinforced glass	Glass with metal wires	Strong, doesn't shatter	Safety windows
MDF (medium density fibre)	Wood fibre and resin	Light, doesn't chip	Furniture

Key words: ceramic, brittle, composite

▶ **2** *Which of the composites listed in the table above would you use to make a playground slide? Why?*

Student Book
pages 120–121

7.7 Building sustainable homes

Wood versus brick

● Many parts of homes can either be made of brick, concrete, polymers or timber (wood). This affects their **carbon footprint**.

● Limestone-based materials like concrete have a big carbon footprint. This is because carbon dioxide is released making cement and quicklime. **Timber homes** have much smaller carbon footprints than **brick homes**. This is because carbon dioxide was absorbed by the wood whilst it was growing, through photosynthesis.

▶ **1** *Why does wood have a low carbon footprint?*

Cob and hempcrete

● **Cob homes** are traditional homes made from soil, clay and straw. These materials need less energy to produce. Cob homes also cost less to heat and cool, because the walls can absorb a lot of energy. This is called **passive solar heating**. **Hempcrete** is similar to cob – it is a mixture of hemp (a plant related to marijuana) and slaked lime.

▶ **2** *What are cob homes made from?*

Straw bales and honeycomb clay blocks

● Straw can be tightly tied into solid blocks and used to build houses. Straw bales are excellent insulators, so it is cheap to heat **straw bale homes**. Straw also has a small carbon footprint.

● **Honeycomb clay blocks** are also good insulators. They last longer than straw bales, but have a bigger carbon footprint.

▶ **3** *Describe **one** advantage and **one** disadvantage of using straw bales instead of honeycomb blocks.*

Key words: carbon footprint, passive solar heating, honeycomb clay block

Key points

● Building conventional brick homes has a larger carbon footprint than using sustainable materials.

● Timber is a sustainable building material that 'locks up' carbon dioxide from when it was a living tree.

● Cob and straw bale houses use local materials and passive heating and cooling.

AQA *Examiner's tip*

Each of these home-building materials has advantages and disadvantages. Make sure you can describe both. Practise by comparing pairs of these materials, such as wood versus hempcrete or brick versus honeycomb clay block.

Bump up your grade

For full marks, remember to **explain** the reasons for choosing a particular material for a job. These reasons should relate to its structure, properties and sustainability.

1 How is limestone changed into quicklime?

2 Write word and symbol equations to show the production of quicklime and slaked lime.

3 Which material is present in concrete but not mortar?

4 Limestone is used to make glass. What other two materials are used?

5 Name three parts of a home made from metal.

6 Ethene (C_2H_4) can be polymerised to form polyethene. Write a symbol equation to show this happening.

7 Why are sinks often made from a ceramic material?

8 MDF is a material made from wood fibres and resin. What type of material is MDF?

9 State one advantage and one disadvantage of building timber houses, compared with brick houses.

10 What are homes made from straw, soil and clay called?

11 Why do straw bale homes have a smaller carbon footprint than brick homes?

12 Why do cob homes cost less to heat and cool?

Chapter checklist ✓ ✓ ✓

Tick when you have:

reviewed it after your lesson	☑ ☐ ☐	
revised once – some questions right	☑ ☑ ☐	
revised twice – all questions right	☑ ☑ ☑	

Move on to another topic when you have all three ticks

Limestone as a building material	☐	☐	☐
Limestone as a starting point	☐	☐	☐
Products of limestone at work	☐	☐	☐
Metals for construction	☐	☐	☐
Polymers in the home	☐	☐	☐
Ceramics and composites in the home	☐	☐	☐
Building sustainable homes	☐	☐	☐

Student Book
pages 124–125

8.1 Everyday fuels

- Many different fuels are used in the UK. Here are the most common:

Fuel	Uses	Why do we use it?
Natural gas (methane)	Domestic boilers, cookers	Can be compressed and stored; doesn't wear out pipes like liquids do
Petrol	Cars, taxis	Petrol engines are quieter and faster than diesel
Diesel	Vans, buses, trucks	Cleaner and much more efficient than petrol
Heating oil	Domestic boilers	Can be stored in tanks where there isn't a permanent gas supply
Kerosene (paraffin)	Jet planes	Lower freezing point than diesel so won't freeze high up in the air
Coal	Heating homes, power stations	Easy to store; contains a lot of energy, doesn't easily catch fire
Propane	Gas barbeques	Easy to compress and store in tanks

⟹ 1 *What might happen if diesel was used to fuel jet planes?*

- Most fuels used in the UK come from **crude oil**. Crude oil is a mixture of compounds called hydrocarbons. **Hydrocarbons** contain carbon and hydrogen only.

Key words: crude oil, hydrocarbon

Student Book
pages 126–127

8.2 Burning fuels

- Fuels react with oxygen from air to release energy. This produces carbon dioxide and water. Burning hydrocarbons in plenty of oxygen is called **complete combustion**. Burning in limited oxygen results in **incomplete combustion**. This produces carbon monoxide and soot as well. It also releases less energy than complete combustion.

⟹ 1 *What are the differences between complete and incomplete combustion?*

- Complete combustion of a hydrocarbon always produces carbon dioxide and water. This can be shown with a word equation:

  methane + oxygen ⟶ carbon dioxide + water

- Another way to show this reaction is with ball-and-stick diagrams, which show how the chemical bonds change:

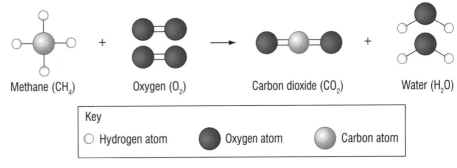

Methane (CH₄)   +   Oxygen (O₂)   →   Carbon dioxide (CO₂)   +   Water (H₂O)

Key
○ Hydrogen atom   ● Oxygen atom   ○ Carbon atom

Complete combustion of methane – ball-and-stick model

8.3 Differences between hydrocarbons

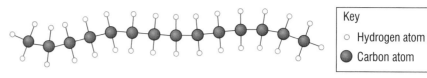

Key	
○	Hydrogen atom
●	Carbon atom

This hydrocarbon has the formula $C_{14}H_{30}$

Key points

- Alkane hydrocarbons have the structure C_nH_{2n+2}.
- Longer chains are darker, thicker and harder to boil and burn.
- The longer the chain, the more oxygen it needs when it combusts.
- We can accurately describe combustion reactions with balanced symbol equations. [H]

- Crude oil is a mixture of hydrocarbons, which are chains of carbon and hydrogen atoms. The length of the chain affects the hydrocarbon's properties.

Short chains	Long chains
Light coloured	Dark coloured
Thin and runny (very short chains are gases)	Thick and viscous
Catch fire easily	Do not catch fire easily

- Longer chains also need more oxygen to burn, as they contain more carbon and hydrogen.

 1 *How many hydrogen atoms are in dodecane, which contains 12 carbon atoms?*

Maths skills

The hydrocarbons on this page all belong to a family called **alkanes**. The number of hydrogen atoms in an alkane molecule is connected to the number of carbon atoms by the formula C_nH_{2n+2}. This means there are always twice plus 2 the number of hydrogen atoms as carbon atoms.
For example, a molecule of decane has 10 carbon atoms. This means it has
$(10 \times 2) + 2 = 22$ hydrogen atoms.

Balancing combustion equations

You need to be able to balance combustion equations.

These steps will help you:

- If there is an even number of carbons in the alkane, start by doubling the amount of alkane.
- Then add more CO_2 to balance the carbon, and more water to balance the hydrogen.
- Finally, add more oxygen to balance the oxygen atoms:

Pentane:

$C_5H_{12} + O_2 \longrightarrow CO_2 + H_2O$ is balanced to $C_5H_{12} + 8O_2 \longrightarrow 5CO_2 + 6H_2O$

Butane:

$C_4H_{10} + O_2 \longrightarrow CO_2 + H_2O$ is balanced to $2C_4H_{10} + 13O_2 \longrightarrow 8CO_2 + 10H_2O$

 2 *Write a balanced symbol equation for the complete combustion of methane (CH_4).*

Key words: alkane

AQA Examiner's tip

Exam questions may ask you to describe or compare the properties of a hydrocarbon. Make sure you relate your answer to the chain length. For example, C_2H_6 is easier to burn than C_5H_{12} because it has a shorter chain length.

8.4 Problems with fossil fuels

Student Book
pages 130–131

Key points

- Fossil fuels are running out. They are non-renewable.
- Burning fossil fuels produces greenhouse gases, which can affect our climate.
- Fossil fuels can harm the environment in other ways, such as acid rain or oil spills.

- Coal reserves should last until the twenty-third century, but oil and natural gas may run out as early as 2050! Fossil fuels are **non-renewable**. Once they're gone, they're gone. Crude oil is used to make many other products as well as fuels, so when it runs out it will have a big impact on us.

▐▐▶ **1** *What does non-renewable mean?*

As well as carbon dioxide, fossil fuels produce other products which harm the environment:

- Sulfur dioxide and nitrogen oxides: These are produced by cars and power stations. They cause acid rain.
- Soot: Particles of carbon are made in incomplete combustion. They can cause breathing problems.
- Carbon monoxide: This is another product of incomplete combustion. It is poisonous.
- Drilling for oil can cause other environmental problems. If a drilling station is damaged, oil can leak out into the environment. Oil spills can also occur when crude oil is transported by sea in giant oil tankers. These can destroy habitats and kill wildlife.

▐▐▶ **2** *How can cars cause acid rain?*

Key words: non-renewable

8.5 Generating electricity

Student Book
pages 132–133

Key points

- Coal, oil and natural gas are fossil fuels, formed over millions of years from the remains of animals and plants.
- Fossil fuels release energy when they burn.
- Several stages take place in a fossil fuel power station before the electricity is generated.

Bump up your grade

You will score more marks for explaining the stages taking place in power stations, rather than just stating them as a list.

Key words: fossil fuel, turbine, generator

- **Fossil fuels** (coal, oil and gas) formed over millions of years from animal and plant remains. Worldwide supplies will run out because we use them faster than they form. They are non-renewable.

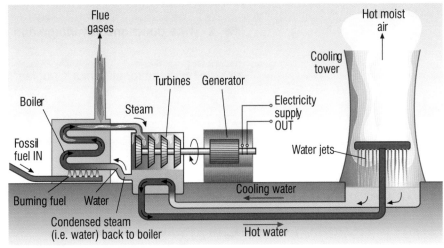

How electricity is generated in a fossil fuel power station

- Burning fossil fuels in power stations changes water to steam. The steam forces a **turbine** and **generator** to spin, generating electricity. Gas-fired power stations spin turbines using the hot exhaust gases from combustion as well.

▐▐▶ **1** *List the stages taking place in a fossil fuel power station.*

Student Book
pages 134–135

8.6 The nuclear alternative

- Nuclear power stations use **uranium** or **plutonium** as a fuel. **Nuclear fission** reactions in the nuclear fuel release energy as heat. This changes water to steam in a separate piped system. Jets of steam force a **turbine** connected to a **generator** to spin. This generates electricity.

> **1** *What stages take place in a nuclear power station?*

Key points

- Nuclear power generates electricity using nuclear fission reactions.
- Nuclear fuels are uranium and plutonium.
- There are advantages and disadvantages in using nuclear power.
- Nuclear power does not produce greenhouse gases but produces radioactive waste.

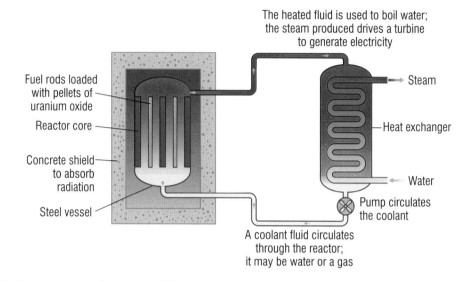

The heated fluid is used to boil water; the steam produced drives a turbine to generate electricity

Fuel rods loaded with pellets of uranium oxide

Reactor core

Concrete shield to absorb radiation

Steel vessel

Steam

Heat exchanger

Water

Pump circulates the coolant

A coolant fluid circulates through the reactor; it may be water or a gas

What happens in a nuclear power station

AQA Examiner's tip

Most processes are the same in fossil fuel power stations and in nuclear power stations, but there are some differences. Learn these carefully so you describe the correct processes in your answer.

There are advantages and disadvantages:

- Nuclear power is reliable, safer than using fossil fuels and does not release greenhouse gases. The fuel is cheap and abundant.
- Nuclear power stations are expensive to build and decommission. Small quantities of highly radioactive waste are produced which must be stored safely for many years.

> **2** *Write down **one** advantage and **one** disadvantage of using nuclear fuels.*

Key words: uranium, plutonium, nuclear fission

Student Book
pages 136–137

8.7 Renewable energy resources

- **Renewable energy** resources will not run out.

▐▐▐➡ **1** *What is meant by a renewable energy source?*

Method	Advantages	Disadvantages
Biomass plants generate electricity when wood, poultry litter and straw are burned.	Reliable Generate large amounts of electricity	Emit greenhouse gases
Blades on **wind turbines** spin in the wind, generating electricity.	Energy source is free No polluting gases produced	Unreliable and do not work in calm weather or storms Create noise pollution
Hydroelectric schemes trap water behind dams, releasing it through pipes to spin turbines when needed.	Reliable Electricity is produced quickly	Only mountainous areas are suitable They disrupt river flows Large areas are flooded
Tidal power schemes trap sea water behind barriers in estuaries, which is released through pipes to spin turbines.	Reliable Large amounts of electricity are generated at predictable times	Flood large areas Expensive to build Only some sites are suitable
Geothermal schemes are used in volcanic regions. Cold water is pumped down deep shafts in the Earth's surface. Heat from rocks, changes the water to steam, which spins turbines in a power station.	No polluting gases produced Energy source is free	Only volcanic areas are suitable Can be expensive to set up
Small floating generators are powered by **wave power**.	No polluting gases produced Energy source is free	Unreliable Produce very small amounts of energy
Solar cells generate small amounts of electricity directly from sunlight.	Solar cells are useful in portable devices No polluting gases are produced	No electricity is produced at night Unreliable Expensive way to generate electricity

▐▐▐➡ **2** *Why do some countries use more renewable energy than others?*

Key words: renewable energy, biomass, wind turbine, hydroelectric, tidal power, geothermal, wave power, solar cell

Key points

- Renewable energy sources will not run out.
- The main renewable energy sources are biomass, wind, hydroelectricity, tidal, geothermal, solar and wave power.
- There are advantages and disadvantages in these schemes.
- Renewable energy sources use free resources and do not produce greenhouse gases. However, they can have an impact on the environment.

Bump up your grade

A good answer will mention factors like the weather, time of day or landscape when considering how well a renewable energy source performs in different situations.

AQA Examiner's tip

To gain extra marks you must be able to describe the environmental and economic impact of using different renewable energy sources.

Student Book
pages 138–139

8.8 The National Grid

- The **National Grid** distributes electricity throughout the country. Pylons support power cables. These transfer electricity from power stations to final users.
- Electricity is generated in power stations. **Step-up transformers** increase the voltage before electricity is transmitted. This allows the same power to be transmitted by a lower current, so there is less energy wastage by heating.
- **Step-down transformers** in **sub-stations** reduce voltages to 230 V, which is safe for final users.

▶ **1** *Write down the parts in the National Grid starting from the power stations and ending in the home.*

▶ **2** *Why is electricity distributed through the National Grid at very high voltages?*

Key points

- Electricity is distributed through the National Grid.
- The National Grid includes power stations, sub-stations, pylons and high-voltage cables.
- Electromagnetic fields surround high-voltage cables.
- There are concerns that high-voltage cables may increase health risks, and damage the environment.

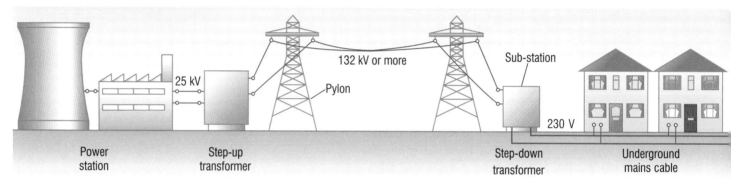

The National Grid

25 kV — Step-up transformer — 132 kV or more — Pylon — Sub-station — 230 V

Power station — Step-up transformer — Step-down transformer — Underground mains cable

Bump up your grade

To get a good mark, you should be able to explain what each part of the National Grid does as well as label a diagram, or put the parts in order.

- Electric and magnetic fields are areas where electric or magnetic forces can be detected.
- These electric and magnetic fields surround high-voltage cables as well as electrical equipment. They are weaker when measured further from the equipment, and when the current or voltages are lower.

Experts do not believe the high voltages from the cables cause harm because:

- Electric fields from the high-voltage cables cannot pass through buildings.
- Magnetic fields from high-voltage cables are much weaker than the Earth's magnetic field.
- Experiments using electric and magnetic fields have not caused changes to living cells.

AQA Examiner's tip

Be prepared to discuss whether the risks from high-voltage cables are acceptable from the points of view of different groups of people.

- Electricity pylons and their high-voltage cables running across the countryside are an eyesore. They spoil some beautiful views. The cables could be run underground but this option is more expensive at the outset of a project.

▶ **3** *Why do scientists believe magnetic fields from high-voltage cables do not cause harm?*

Key words: National Grid, step-up transformer, step-down transformer, sub-station

1 What are the products when a hydrocarbon fuel is burned with plenty of oxygen?

2 Write a word equation for the complete combustion of hexane.

3 Write a balanced symbol equation for the combustion of propane (C_3H_8). [H]

4 Name two products of incomplete combustion that are not produced in complete combustion.

5 Fill in the number of hydrogen atoms in these alkanes: $C_3H_?$, $C_{11}H_?$, $C_9H_?$, $C_8H_?$

6 List all the ways that extracting and using crude oil can damage the environment.

7 Describe three ways that water is used to generate electricity.

8 Write down a flow chart showing how electricity is generated from nuclear fuels.

9 Why are fossil fuels described as non-renewable?

10 Explain why wind power is more unreliable than hydroelectricity.

11 Write down two environmental problems caused by hydroelectricity.

Chapter checklist	✓	✓	✓
Tick when you have:			
reviewed it after your lesson ✓ ☐ ☐	Everyday fuels	☐	☐	☐
revised once – some questions right ✓ ✓ ☐	Burning fuels	☐	☐	☐
revised twice – all questions right ✓ ✓ ✓	Differences between hydrocarbons	☐	☐	☐
Move on to another topic when you have all three ticks	Problems with fossil fuels	☐	☐	☐
	Generating electricity	☐	☐	☐
	The nuclear alternative	☐	☐	☐
	Renewable energy resources	☐	☐	☐
	The National Grid	☐	☐	☐

9.1 Energy

Student Book
pages 144–145

Key points

- Energy is measured in joules, J, or kilojoules, kJ.
- Sankey diagrams show the energy transfers taking place.

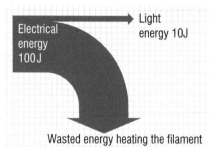

This light bulb transfers 100 J of electrical energy to 90 J of wasted energy and 10 J of light energy

- Energy makes it possible for things to happen. Energy is measured in **joules (J)** or **kilojoules (kJ)**.
- 1 kilojoule is 1000 J
- When energy is transferred, no energy is lost or created but there are unwanted transfers to the surroundings (wasted energy).
- The total energy transferred by a device matches the amount of energy supplied.

 1 *What happens to the total amount of energy during an energy transfer?*

- **Sankey diagrams** show the energy flow through a device.
- The input form of energy is shown on the left-hand side of the arrow.
- Each output form of energy is shown as a separate arrow.
- The width of each section shows the proportion of each type of energy.

Maths skills

Read the diagram from left to right. Unwanted forms of energy are usually shown at the bottom of the diagram.

AQA Examiner's tip

You could be asked to draw a Sankey diagram, or to describe the energy changes that it shows.

Key words: joule (J), Sankey diagram

9.2 Electrical power

Student Book
pages 146–147

Key points

- Power is the rate at which energy is transferred.
- Power is measured in watts, W, or kilowatts, kW.
- Power is calculated using: power = energy ÷ time, or power = voltage × current

Bump up your grade

If the question does not show the unit, then an extra mark is available for including the correct unit in your answer.

- **Power** measures how quickly energy is transferred. It is measured in **watts** or **kilowatts**. A kilowatt is 1000 watts. If the power of a device is one watt, then it transfers one joule per second.
- Another name for **potential difference** is voltage. In the UK, mains electricity is supplied at 230 V.

 1 *What is the power of a hairdryer that transfers 3.6 kJ in 3 seconds?*

2 *What is the power of a vacuum cleaner if 4 A flows when the supply voltage is 230 V?*

Maths skills

Measuring power

We calculate electrical power using one of these equations:

power (W) = potential difference (V) × current (A)

power (W) = energy transferred (J) ÷ time (s)

To decide which equation to use, check what information is given in the question.

Key words: power, watt, potential difference

Student Book
pages 148–149

Key points

- Electricity is sold in Units of kilowatt-hours (kWh).
- Units are calculated using power (kW) × time (h).
- Electricity meter readings are used to show the kilowatt-hours used.
- Total cost = kilowatt-hours used × cost per kilowatt-hour.

9.3 Buying electricity

- A **kilowatt-hour (kWh)** is the energy used if a 1 kW device is switched on for one hour.
- A kWh is also called a **Unit**.
- Calculate kilowatt-hours (kWh) using power (kW) × time (h).

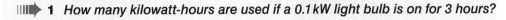 **1** *How many kilowatt-hours are used if a 0.1 kW light bulb is on for 3 hours?*

- The electricity meter records how many kWh are used at any time. An electricity bill is calculated in kWh by multiplying the number of kWh used since the last reading by the cost per kWh.

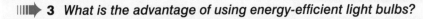 **2** *The reading on an electricity meter is 03125. Three months ago it was 03075. How many Units of electricity were used?*

You can reduce your electricity bills by:
- using more efficient equipment
- switching on equipment for a shorter time
- paying a lower price per Unit.

3 *What is the advantage of using energy-efficient light bulbs?*

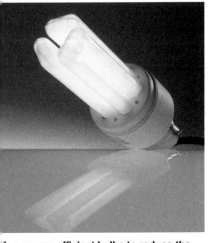

Use energy-efficient bulbs to reduce the electricity used compared with filament bulbs

Bump up your grade

Make sure you show the numbers you use to calculate an answer. You will get marks for the right method even if you make a mistake converting power or time.

AQA Examiner's tip

Make sure you know if your answer was calculated in pounds or in pence.

 ### Maths skills

Divide power in watts by 1000 to change the unit of power to kilowatts, kW.
Divide the time in minutes by 60 to change times into hours.

Key words: kilowatt-hour (kWh), Unit

Student Book
pages 150–151

9.4 Efficiency

- **Efficiency** measures the proportion of input energy that is usefully transferred by an **appliance**. Calculate efficiency using:

$$\text{Efficiency} = \frac{\text{useful energy out}}{\text{total energy in}},$$

or $$\text{Efficiency} = \frac{\text{useful power out}}{\text{total power in}}.$$

 1 *What is the efficiency of a motor that provides 3000 J of useful output when 6000 J of energy is supplied?*

- The efficiency of electrical equipment like fridges and washing machines is shown on an **energy label**. More efficient equipment wastes less energy and costs less to run. Most energy is wasted heating the device and its surroundings, or as sound.

Maths skills

Calculating efficiency as a percentage
An electric heater produces 190 J of useful output energy for every 200 J of electrical input energy.
Its efficiency is (190 ÷ 200) × 100% = 95%

Key words: efficiency, appliance, energy label

Key points

- Efficiency measures how much energy is usefully transferred.
- Efficiency can never be more than 100%.
- Efficiency is calculated as useful energy out ÷ total energy in.
- Efficiency can also be calculated as useful power out ÷ total power in.
- Energy labels on appliances indicate the efficiency.

AQA Examiner's tip

Efficiency can never be more than 1.0 or 100%. Check your calculation if this happens.

Student Book
pages 152–153

9.5 How fast do waves travel?

- Waves are regular disturbances that transfer energy from one place to another, without transferring matter.
- The number of waves produced each second is its frequency, measured in **hertz**. One hertz (Hz) is a frequency of one cycle per second.
- The velocity of a wave is calculated using:

velocity (m/s) = **frequency** (Hz) × **wavelength** (m)

 1 *A wave has a frequency of 500 Hz and its wavelength is 0.2 m. What is the speed of this wave?*

Maths skills

Wave calculations
Since electromagnetic waves travel at 300 million m/s:
- the frequency of electromagnetic waves is 300 million ÷ wavelength
- the wavelength of electromagnetic waves is 300 million ÷ frequency.

2 *A microwave has a wavelength of 3 cm, and its velocity is 300 million m/s. What is its frequency?*

Key points

- All waves transfer energy without transferring matter.
- Waves are regular disturbances passing through matter.
- The number of waves produced each second is the frequency, measured in hertz (Hz).
- Velocity = frequency × wavelength.

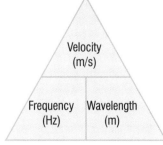

Use this triangle to calculate a wave's velocity, frequency and wavelength

Key words: velocity, frequency, wavelength

Student Book
pages 154–155

9.6 Electromagnetic waves

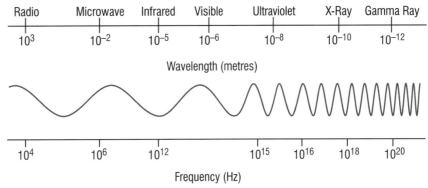

The electromagnetic spectrum

Key points

- The energy of electromagnetic waves increases with frequency.
- Radio waves are used for TV and radio broadcasts.
- Microwaves are used for mobile phone networks, satellite TV and cooking food.
- Infrared waves are used in TV and DVD remote controls.
- Visible light is used in fibre optic cables.
- Ultraviolet is used in sun beds.

- **Electromagnetic waves** all travel through space at the speed of light, 300 million m/s. Higher frequency waves carry more energy. In order of increasing energy, the order of the electromagnetic spectrum is:

  Radio waves, microwaves, infrared radiation, visible light, ultraviolet radiation, X-rays, gamma rays

- Radio waves transmit TV and radio broadcasts.
- Microwaves transmit satellite TV broadcasts and mobile phone signals. Microwaves are used for cooking because food that absorbs microwaves gets hotter.
- Infrared wavelengths are used in TV remote controls and fibre optic communication.
- Visible light travels inside **fibre optic cables**. These are very fine strands of glass. Fibre optic cables are used in internet and telephone connections and cable TV. In hospitals, **endoscopes** use fibre optic cables to look inside the body.
- Skin that absorbs **ultraviolet radiation** becomes tanned. Sun beds work by producing ultraviolet radiation.

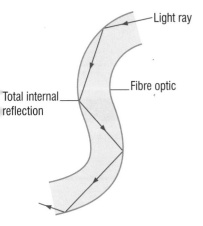

Light ray

Fibre optic

Total internal reflection

Optic fibres carry many messages simultaneously

> 1 Which type of radiation is used:
>   a to cook food
>   b to produce a suntan
>   c in TV remote controls?

Key words: electromagnetic wave, radio wave, microwave, infrared radiation, visible light, ultraviolet radiation, X-ray, gamma ray, fibre optic cable, endoscope

AQA Examiner's tip

Make sure you know which properties are shared by all members of the electromagnetic spectrum.

**Student Book
pages 156–157**

9.7 Dangers of radiation

Key points

- Microwaves and ultraviolet radiation can be harmful.
- Hospitals use X-rays to produce shadow pictures of bones and gamma rays are used to treat cancer.
- X-rays and gamma rays are harmful in large doses so their use is controlled.

- X-rays and gamma rays have a short wavelength and high frequency. They carry enough energy to kill or damage cells if they are absorbed, so they are not normally used in the home.
- X-rays pass through soft tissues, but are absorbed by bone and teeth. Hospitals and dentists use X-rays to produce shadow pictures.
- In high doses, gamma rays kill or damage cells making them cancerous. Gamma rays are used to treat cancer by killing cancer cells.
- Ultraviolet radiation causes sunburn within minutes of exposure to intense radiation and can cause skin cancer with over-exposure.
- Microwaves are absorbed by cells, heating them. High doses can damage cells, but low doses are unlikely to cause harm.

▕▋▋▋➤ **1** *Why are gamma rays more damaging than ultraviolet radiation?*

▕▋▋▋➤ **2** *Why are X-rays and gamma rays only used in controlled settings?*

Bump up your grade

To get the best marks you should be able to link the uses or hazards of a particular type of wave to the frequency and energy of the radiation being discussed.

AQA Examiner's tip

Gamma radiation causes cancer by damaging cells, but also treats cancer by killing cancer cells.

X-rays and gamma rays can only be used by authorised workers in controlled areas

1 What units are used to measure energy?

2 The diagram shows the energy transfers taking place in a motor. Use the diagram to answer the questions:

  a Write down the energy transfers shown in the diagram.

  b How much energy is changed into heat and sound energy?

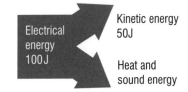

3 How are power and energy linked?

4 **a** What is the power of a bulb that transfers 50 joules per second?

  b How much energy is transferred by the bulb in five minutes?

5 What is the power of a radio if it uses four 1.5 V batteries (a total of 6 V) and a current of 0.2 A flows through it?

6 An electric lift converts 10 000 J of electrical energy into 7500 J of gravitational potential energy. What is its efficiency?

7 What is the velocity of a wave that has a wavelength of 20 m and a frequency of 6 Hz?

8 List the members of the electromagnetic spectrum in order, starting with the highest frequency.

9 What do all electromagnetic waves have in common?

10 Write down two uses of microwaves.

11 Why is the use of gamma rays and X-rays controlled?

Chapter checklist ✓✓✓

Tick when you have:				Energy	☐	☐	☐
reviewed it after your lesson	✓	☐	☐	Electrical power	☐	☐	☐
revised once – some questions right	✓	✓	☐	Buying electricity	☐	☐	☐
revised twice – all questions right	✓	✓	✓	Efficiency	☐	☐	☐
Move on to another topic when you have all three ticks				How fast do waves travel?	☐	☐	☐
				Electromagnetic waves	☐	☐	☐
				Dangers of radiation	☐	☐	☐

1 Some diseases can be inherited. Cystic fibrosis is a genetically inherited disease that is on a recessive gene.

  a Draw a Punnett square diagram to help you find the probability of a child inheriting cystic fibrosis when only one parent has the disease and the other parent is a carrier. Use F for the healthy gene and f for the diseased gene. *(4 marks)*

  b Explain how two healthy parents could give birth to a child with sickle-cell anaemia. *(2 marks)*

2 To stay healthy, the body must keep itself at the right temperature.

  a Explain how blood vessels can help decrease the temperature of the body. *(3 marks)*

  b Explain how sweating can help decrease the temperature of the body. *(2 marks)*

  c Describe how the body maintains a steady temperature through negative feedback. **[H]** *(3 marks)*

3 Farmers use slaked lime on their fields as it neutralises the soil and raises the pH.

  a What is a neutralisation reaction? *(2 marks)*

  b What is the chemical formula for limestone? *(1 mark)*

  c To make slaked lime, limestone is converted to quicklime in a lime kiln. Describe how converting limestone into quicklime affects the environment. *(2 marks)*

  d The quicklime can then be converted into slaked lime by adding water. This reaction is an exothermic reaction.

   i Copy and complete the chemical equation of the conversion of quicklime to slaked lime:

   $$... + H_2O \longrightarrow ...$$ *(2 marks)*

   ii What is an exothermic reaction? *(1 mark)*

4 Hydrocarbons can be used as fuels to heat homes and for transport.

  a What is a hydrocarbon? *(1 mark)*

  b Copy and complete the balanced symbol equation for the complete combustion of propane.

   $$C_3H_8 + ... \longrightarrow ... + ...$$ **[H]** *(3 marks)*

5 A torch uses a voltage of 6 V and current of 2 A supplied by a battery.

  a Label the Sankey diagram for a torch with the correct types of energy.

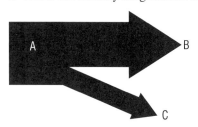

(3 marks)

  b Calculate the power used by the torch using the equation:

   power = voltage × current *(2 marks)*

  c Calculate the efficiency of the torch if 9 W is given off as useful power. *(2 marks)*

6 The table shows a comparison of the electrical conductivity of different materials. Conductivity is measured in Siemens per metre where the higher the number the better the electrical conductor. Use the data in the table to answer the questions.

Material	Conductivity in Siemens per metre
Silver	63.0×10^6
Copper	59.6×10^6
Gold	45.2×10^6
Sea water	4.8
Drinking water	0.0005 to 0.05

a Give **one** advantage and **one** disadvantage of using copper compared with the other materials in the table for electrical wiring. *(2 marks)*

b Gold is often used in electrical cabling because gold does not tarnish. Calculate the percentage loss in conductivity compared with using a silver cable. *(3 marks)*

7 *In this question you will be assessed on using good English, organising information clearly and using specialist terms where appropriate.*

Some people are worried that mobile phones are hazardous to health. The table gives their opinions about mobile phones and the arguments against those opinions.

Mobile phones are dangerous	Mobile phones are safe
The waves given off by mobile phones can heat up body tissues.	The waves given off by mobile phones are not powerful enough to cause damage to the body.
People who make long phone calls on their mobile often complain of tiredness, headaches and loss of concentration.	Results have never been reported in laboratory conditions that tiredness, headaches and loss of concentration are due to mobile phones so these may be due to other factors including lifestyle.
Mobile phone users are 2.5 times more likely to develop cancer in areas of the brain adjacent to the ear they normally use for their phone.	Researchers admit that there is no clear connection between brain cancer and mobile phone use.
The Internal Agency for Research on Cancer found a link between childhood cancer and living next to mobile phone masts.	Modern mobile phone masts release very low levels of radiation.

Use the information in the table to help you explain why the waves used in long-distance mobile phone use are safe. *(6 marks)*

Student Book pages 164–165

10.1 Medicines

Medical drugs are legal drugs which cure, prevent or treat disease. Some drugs, i.e. **antibiotics**, kill the microorganism (**pathogen**) that has made you ill.

Other drugs relieve symptoms but do not cure diseases. Examples include:

- Aspirin is an **analgesic** drug that reduces pain. Aspirin is also an **anti-inflammatory** drug that reduces swelling, which in turn reduces pain.
- When you hurt yourself, the body releases a hormone-like substance that makes you feel pain. Paracetamol reduces the production of this chemical, decreasing the pain that you feel.
- High blood pressure treatments reduce the amount of water in the blood and widen arteries, allowing blood to flow more freely. This temporarily reduces strain on the heart and blood vessels.

1 *How do anti-inflammatory drugs work?*

- Long-term use of 'over the counter' drugs or an overdose can lead to serious side effects or even death. The over-use of medical drugs can cause kidney failure and liver damage. Anti-inflammatory drugs can also cause stomach ulcers, or problems with intestines.

Key words: antibiotic, pathogen, analgesic, anti-inflammatory

Key points

- A medical drug improves health by curing, preventing or relieving the symptoms of a disease.
- Analgesic drugs reduce pain. Anti-inflammatory drugs reduce swelling, which may relieve pain.
- Over-use of symptom-relieving drugs may lead to addiction, and kidney and liver damage.

AQA Examiner's tip

Make sure that you know what anti-inflammatory, analgesic and antibiotic drugs do, and are able to name some examples of medical conditions that can be treated by each one.

Student Book pages 166–167

10.2 Antibiotics

- Antibiotics are drugs that kill *bacteria*. They have *no* effect on viruses or many fungi. There are several different types of antibiotic. Each kills a different species, or range of species, of bacteria. One example is penicillin. It is used to treat ear, nose and throat infections.

1 *Which type of microorganisms do antibiotic drugs kill?*

- Bacteria can spontaneously **mutate**, and these mutations can lead to some strains becoming **resistant** to antibiotics. This means that many types of antibiotic will no longer kill them. These **antibiotic-resistant bacteria** are often referred to as 'super bugs'. An example is MRSA.

When antibiotics are used to treat an infection, they will kill individual pathogens that do not have antibiotic resistance. However, resistant pathogens will survive. These will then reproduce, increasing the population of the resistant strain. To try to slow down the rate of development of resistant strains, doctors no longer prescribe antibiotics for non-serious infections, such as mild throat infections.

2 *Name an antibiotic-resistant species of bacteria.*

- In the past, doctors often prescribed antibiotics routinely to treat minor illnesses, i.e. coughs and colds. However, many of these conditions are caused by viruses, so the antibiotics had no effect.

Key points

- Antibiotics are drugs that kill bacteria.
- Over-prescribing antibiotics increased the number of antibiotic-resistant bacteria causing significant costs for the NHS.
- Spontaneous mutations can result in a bacteria developing antibiotic resistance. Antibiotics kill the non-resistant strains, but the resistant strains survive and reproduce, increasing their population. [H]

- The widespread use of antibiotics has increased the rate of development of antibiotic-resistant bacteria. These bacteria survive and reproduce, causing an increase in their numbers. This has created extra costs for the NHS. For example, more members of staff are required to control and treat outbreaks of these infections. Research into developing new drugs and their production is also expensive.

Key words: mutate, antibiotic-resistant bacteria

Student Book
pages 168–169

Key points

- Drugs are tested in the laboratory, on animals and on human volunteers to make sure they are safe.
- Arguments for testing drugs on animals include: saves human lives; shorter life cycles of animals; cheaper than testing on humans.
- Arguments against testing drugs on animals include: the animals suffer; animals have the right to life; results may not be the same for humans.

10.3 Approving a new drug

- Medical drugs have to be tested to ensure they are safe and effective before they can be prescribed by doctors. See the flow chart below:

Drug is tested using computer models and human cells grown in the laboratory. Many drugs fail at this stage because they damage cells or appear not to work.

↓

Drug is tested on animals (nematode worms/fruit flies/mice) to study any side effects.

↓

Drug is tested on a small group of healthy human volunteers to check its safety. Testing drugs on humans is known as clinical trials.

↓

Drug tested on volunteer patients who have the illness, to ensure it works.

↓

Drug tested on patients to monitor drug effectiveness, safety, dosage and side effects.

↓

Drug approved and can be prescribed.

⁍ 1 What are clinical trials?

Arguments for testing drugs on animals:

- Shows how drug affects a living body.
- Animal lives are not as valued as human lives.
- Animals have a shorter lifecycle, so long-term effects can be studied in a relatively short time.
- Many animals can be tested at one time, and it is cheaper than carrying out research on humans.

Arguments against testing drugs on animals:

- Reaction to drug may be different from a human.
- Animals have the right to life.
- Can cause pain or discomfort for the animal.
- Many animals die during the testing, or have to be put down after the trial.

⁍ 2 Give one argument for and one against testing drugs on animals.

AQA Examiner's tip

If you have to describe a process in an exam, such as how a drug is tested to ensure it is safe and effective, try drawing a flow diagram or bulleted list showing the steps involved.

Key words: clinical trial

**Student Book
pages 170–171**

10.4 Recreational drugs

- **Drugs** alter the chemical reactions that take place inside the body. If the body gets used to these changes, it may become dependent on a drug. The person is then **addicted**.
- If an addict attempts to stop taking a drug, they suffer **withdrawal symptoms**. Their body is no longer being provided with a chemical it is used to having. Withdrawal symptoms include headaches, sweating, sickness, insomnia and depression.
- A recreational drug is taken for a person's enjoyment. It has no medical purpose.

> 1 *How do drugs affect the body?*
> 2 *What happens if an addict stops taking a drug?*

Some drugs that can harm our body are legal. These include:
- Alcohol – This affects your nervous system and damages your liver.
- Tobacco – Smoking seriously increases your risk of cancer, and respiratory and heart diseases.
- Antidepressants (prescribed by doctors to relieve depression) – These provide short-term benefits, but can result in addiction.
- Barbiturates (sedatives or tranquillisers, present in some prescribed sleeping tablets) – These are highly addictive and long-term use can lead to depression.

Illegal drugs can damage your body, even in very small amounts. They increase the risk of some lung and heart diseases. They can be grouped together by their effect on the body:
- Stimulants make you feel more alert, awake, and generally happier. They work by speeding up the nervous system. Examples include amphetamines and cocaine.
- Depressants reduce feelings of stress and panic and make you feel more relaxed. They work by slowing down the activity of the nervous system. Examples include barbiturates and heroin.
- Hallucinogens interfere with normal brain function, altering what people see and hear. For example, cannabis.

Key words: drugs, addicted, withdrawal symptoms

Key points

- Drugs are chemicals that affect the body in a helpful or harmful way.
- People can become addicted to drugs and suffer withdrawal symptoms if they try to stop taking them.
- Recreational drugs are taken for personal enjoyment. They have no medical benefits.

**Student Book
pages 172–173**

10.5 Tobacco

- Tobacco smoke contains over a thousand chemicals, many of which are harmful. They include:
  - Tar is a sticky black material, which collects in the lungs and irritates and narrows airways. Some of the chemicals it contains cause cancer.
  - **Nicotine** is an addictive drug which affects the nervous system, makes the heart beat faster, and narrows blood vessels.
  - **Carbon monoxide** is a poisonous gas which reduces the blood's oxygen-carrying capacity. Oxygen is transported by binding to haemoglobin, in red blood cells. But if carbon monoxide is present, this will bind to haemoglobin in preference to oxygen.

> 1 *What does nicotine do?*

Key points

- Tobacco smoke contains tar, nicotine and carbon monoxide.
- Smoking increases your risk of circulatory and respiratory disease.
- Carbon monoxide binds to haemoglobin in red blood cells limiting the oxygen flow around the body.

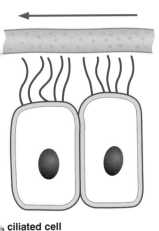

Mucus

ciliated cell

- Smokers' arteries narrow as fatty deposits are left on artery walls. This prevents blood flowing properly. Smokers are also at a higher risk of blood clots, leading to a heart attack or stroke.
- Ciliated cells lining your windpipe sweep mucus (containing trapped dirt and microorganisms) out of the airways and into your stomach. This keeps your airways clean. Chemicals in smoke paralyse the cilia resulting in mucus flowing into the lungs. This makes it hard to breathe and can cause infection. Smokers have to cough this mucus up, causing further lung damage.

2 *Why are smokers at a higher risk of blood clots?*

Bump up your grade

When you are asked to explain how smoking affects the body, you should not only list the health problems caused by smoking, but also explain how the problem is caused.

Key words: nicotine, carbon monoxide

Student Book
pages 174–175

Key points

- Alcohol is a depressant, which acts on the nervous system. It slows down the body's reactions.
- Long-term alcohol consumption can result in brain and liver damage.

10.6 Alcohol

- Alcohol contains the drug **ethanol**. It is absorbed into the bloodstream and travels to the brain, where it affects the nervous system. Ethanol is a depressant – it slows down body reactions and can change behaviour. Most people feel relaxed and happy, but some become aggressive or depressed.

1 *What does alcohol do to the body?*

- Over time, heavy drinking can cause stomach ulcers, heart disease, and brain and liver damage. The liver breaks down ethanol (which is poisonous) into harmless waste products. Heavy drinkers may suffer from cirrhosis of the liver. This means that the liver takes much longer to break down alcohol and other toxins. This can be fatal.
- Binge drinking (consuming large quantities of alcohol in a short time) can result in alcohol poisoning, which can also be fatal. It also increases the risk of long-term health problems.

2 *Name three examples of medical problems caused by heavy drinking.*

- When people drink alcohol regularly, they need more and more to have the same effect on their body. This is because their body has developed a tolerance to ethanol. If they carry on drinking, they may become addicted – an alcoholic.

3 *What is a person called if they are addicted to alcohol?*

Key words: ethanol

1 What is the difference between a medical and a recreational drug?

2 What type of drug is: **a** penicillin **b** paracetamol **c** aspirin?

3 Why are antibiotics not used to treat chicken pox, which is a disease caused by a virus?

4 State two problems of binge drinking.

5 What is the difference between a stimulant and a depressant?

6 Before a drug can be prescribed it has to undergo clinical trials. Reorganise the steps A to F into the correct order:

A Drug is tested on a wide range of people.

B Drug is tested on animals.

C Drug is tested using computer models and human cells.

D Drug is approved and can be prescribed.

E Drug is tested on a small group of healthy human volunteers.

F Drug is tested on volunteer patients who have the illness.

7 **a** What is the drug contained in alcohol?

b How does alcohol affect the body?

8 **a** What is meant by the term 'addict'?

b Why do addicts suffer from withdrawal symptoms if they stop taking a drug?

9 Explain how carbon monoxide in tobacco smoke harms the body.

10 **a** How do antibiotic-resistant strains of bacteria develop? [H]

b Why is the number of antibiotic-resistant strains of bacteria increasing?

Chapter checklist		✓ ✓ ✓
Tick when you have:	Medicines	☐ ☐ ☐
reviewed it after your lesson ☑ ☐ ☐	Antibiotics	☐ ☐ ☐
revised once – some questions right ☑ ☑ ☐	Approving a drug	☐ ☐ ☐
revised twice – all questions right ☑ ☑ ☑	Recreational drugs	☐ ☐ ☐
Move on to another topic when you have all three ticks	Tobacco	☐ ☐ ☐
	Alcohol	☐ ☐ ☐

Student Book
pages 178–179

11.1 Harmful microorganisms

- Microorganisms are living things, at least 100 times smaller than your cells. Most cause no harm to animals or plants. However, some can cause disease when they enter the body. These are called **pathogens**. There is a time delay between pathogens entering the body, and a person feeling unwell. This is called the **incubation period**. During this time, the microorganisms are rapidly reproducing.

> **1** *What is the name given to a microorganism which causes disease?*

> **2** *What is an incubation period?*

Key points

- Bacteria, fungi and viruses are three groups of microorganism.
- Pathogens are microorganisms which cause disease in a plant or an animal.
- Bacteria cause disease by damaging cells or producing toxins. Viruses cause disease by damaging cells.

Bump up your grade

You may be asked to read information from a microbial growth chart. When you take a reading from a graph, use a pencil and ruler to read off the numbers. This makes it clear to the examiner how you reached your answer – see graph.

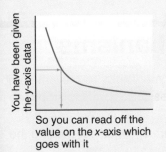

You have been given the *y*-axis data

So you can read off the value on the *x*-axis which goes with it

Microorganism	Bacterium	Virus
Appearance		
Features	• have a cell wall and cell membrane • no nucleus • genetic material floats around in the cytoplasm • larger than a virus	• have a protein coat • no nucleus • a few genes that float around inside the virus • smaller than bacteria
Examples of diseases	• cholera • typhoid • tuberculosis	• measles • rubella • mumps • polio
How they replicate	Bacteria replicate by dividing in half. So 1 bacterium becomes 2 bacteria, then 4, then 8, and so on.	Viruses replicate by taking over the nucleus of the host cell, and instructing the host cell to make copies of the virus. The cell then bursts, releasing the viruses into the body

> **3** *How can you tell the difference between a bacterium and a virus?*

Key words: pathogen

Student Book
pages 180–181

11.2 How are diseases spread?

To cause harm, pathogens have to enter our bodies. This can happen through:

- cuts in the skin – from injury, or insect / animal bites
- the digestive system – when you eat and drink
- the respiratory system – when you breathe through your mouth and nose
- the reproductive system – during sexual intercourse.

You are more likely to become unwell if large numbers of microorganisms enter your body. This can occur through contact with an infected person or exposure to unhygienic conditions.

How can you prevent diseases spreading from an infected person?

- Colds and flu – caused by tiny drops of liquid (**droplet infection**) released during coughing and sneezing can be prevented by covering mouth and nose, e.g. using a mask or handkerchief.
- Contagious diseases like mumps / chicken pox that are spread by touching infected people or contaminated objects can be prevented by not touching infected people or contaminated objects.
- Diseases like syphilis, gonorrhoea and HIV that are spread through sexual intercourse can be prevented by using condoms.
- Diseases that are spread through the blood like HIV and hepatitis can be prevented by using new, sterilised needles.

You can prevent diseases spreading by being hygienic in the following ways:

- Cooking food properly – this kills any microorganisms present.
- Maintaining good personal hygiene – prevents the transmission of microorganisms.
- Drinking clean treated water – water sterilisation kills any microorganisms present.
- Covering cuts and grazes – this prevents infection as microorganisms cannot enter your body.

▐▐▶ 1 *Name three ways pathogens can enter your body and three ways you can hygienically protect against their spread.*

Key points

- Pathogens can enter the body through wounds, during sexual intercourse, or through the digestive and respiratory systems.
- Contact with infected people, or unhygienic conditions can spread diseases.
- Avoiding infected people and being hygienic can prevent diseases spreading.

AQA Examiner's tip

Make sure you read the question properly. If you are asked to name two ways you can prevent a disease being spread, make sure you do not give the same answer twice. For example 'cover mouth when sneezing' and 'cover mouth when coughing' are the same as both prevent droplet infection.

Key words: droplet infection

Student Book
pages 182–183

11.3 Body defence mechanisms

- If your skin is cut or grazed, pathogens can enter your body. To prevent pathogen entry, the skin needs to seal a cut as quickly as possible. This also stops too much blood being lost.
- **Platelets** in blood are small pieces of cell which are essential for helping the blood to clot.

Key points

- Platelets help the blood to clot by forming scabs.
- Phagocytes and lymphocytes are white blood cells.
- Antibodies (chemicals) 'deactivate' microorganisms, stopping them from causing disease.

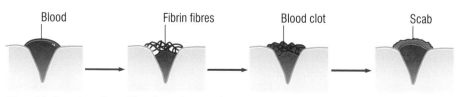

Blood   Fibrin fibres   Blood clot   Scab

The skin is cut and starts to bleed – blood leaks out of the body.	Platelets change the blood protein fibrinogen into fibrin. This forms a network of fibres in the cut.	Red blood cells are trapped in the fibres. This forms a blood clot.	The clot hardens to form a scab. This keeps the skin clean and gives it time to heal. In time, the scab falls off.

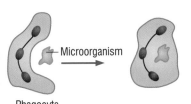

Phagocyte

Phagocyte engulfing a pathogen

> **1** *What do platelets do?*

Sometimes, harmful microorganisms do manage to enter our bodies. It is the job of white blood cells to prevent them causing disease. There are two types of white blood cells:

- **Phagocytes** are cells that engulf microorganisms and make enzymes which break down the microorganism.
- **Lymphocytes** are cells that make **antibodies** (a chemical). An antibody reacts with the microorganism and de-activates it. Each antibody is specific for one type of microorganism. Lymphocytes also make **anti-toxins**. These chemicals destroy the poisonous toxins that some microorganisms make.

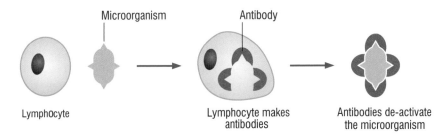

Lymphocyte fighting disease

> **2** *How do lymphocytes fight disease?*

Key words: platelet, antibody, anti-toxin

Student Book
pages 184–185

Key points

- Once antibodies have fought off a disease, some remain in your body. This gives you immunity to the disease.
- Immunity prevents you suffering from the same disease again.
- Immunisations trigger your body to produce antibodies against a dead or weakened microorganism. This gives you protection against a disease.

11.4 Immunisation

- When a pathogen enters your body, lymphocytes make an antibody against it. After the antibodies have 'fought off' the disease, some remain in the body. If the same type of microorganism enters your body again, the antibodies will destroy it before it can cause disease. This is called **immunity**. It prevents you suffering from the same disease again.

> **1** *What does it mean if a person is immune to a disease?*

- **Immunisations** (also referred to as vaccinations) can protect you against some diseases caused by microorganisms. For example, you can receive immunisations against polio, measles and TB.
- Immunisation involves a **vaccine** being inserted (normally through an injection) into your body. Most vaccines contain dead or weakened microorganisms that can no longer cause disease. The microorganisms do not make you ill, but still trigger your lymphocytes to make antibodies. The antibodies destroy the microorganisms. Some remain in your body to fight off the pathogen quickly if it enters your body again, preventing it causing disease. You are now immune.

> **2** *What does a vaccine normally contain?*

Key words: immunisation, vaccine

Student Book
pages 186–187

11.5 Vaccination issues

- Scientific data provides evidence that immunisations work. For example, between 1951 and 2005 measles injections were given to large numbers of the UK population. Despite a rise in population during this time, the number of cases of measles decreased.

Immunisations protect against many life-threatening conditions. However, not everyone chooses to be immunised. This may be because of:

- Occasional scares about the safety of some vaccines. However, vaccines have to undergo many stages of testing to ensure their safety.
- Possible side effects. Most people suffer no side effects, and any experienced are normally mild and very short-lived. Side effects may include fever, sickness, diarrhoea, swollen glands and irritability. Severe reactions are very rare.
- Concerns that vaccines overload our immune system, making it less able to react to other diseases such as meningitis, AIDS and cancer. However, there is no evidence to suggest this is true. Scientific studies estimate that vaccines occupy less than 0.1% of our immune systems.

Key points

- Scientific data provides evidence that immunisations have decreased the number of cases of infectious diseases.
- Most people suffer no (or very mild) side effects from immunisation.
- Some people choose not to have immunisations because they are worried about side effects.

➤ **1** *How is the safety of vaccines ensured?*

➤ **2** *Name three possible side effects of an immunisation.*

Bump up your grade

If a question on immunisation effectiveness contains a graph that you need to interpret, then make sure that you:
- State any trends that can be seen in the data.
- Use data from the graph to support your answers.

Student Book
pages 188–189

11.6 Medical uses of X-rays

- **X-rays** (a form of **ionising radiation**) are high-energy **transverse** waves. They are just a small part of a large family of waves known as the electromagnetic spectrum.
- X-rays are used to see inside the body, without an operation. This removes the chance of infection. X-ray images can diagnose many conditions including broken bones, tumours and chest infections.

➤ **1** *Name three conditions that X-rays can diagnose.*

X-ray images are produced by:
1. Placing photographic film behind the patient, with an X-ray generator in front.
2. The X-rays penetrate through soft tissues like skin and muscle. They are absorbed by denser structures, such as bones and teeth.
3. The X-rays that penetrate through the patient expose photographic film. These areas appear black on the developed film. Areas not exposed to X-rays appear white.

- Radiographers are at risk of receiving high doses of radiation, which could cause cancer. For protection, a lead screen (which absorbs X-rays) is placed between the radiographer and the patient. Radiographers also wear **film badges** which measure the radiation dose received.

➤ **2** *How do radiographers monitor their exposure to X-radiation and protect themselves from the X-rays?*

Key points

- X-rays are high-energy, transverse, electromagnetic waves.
- 'Shadow pictures' are images of the inside of the body, produced using X-rays.
- Radiographers and medical professionals need to be protected from the harmful effects of X-rays.

Bump up your grade

If you are asked to explain how an X-ray image is made, use a bullet-pointed list, or flow chart. Make sure the number of steps in your answer matches the number of marks available.

Key words: X-ray, ionising radiation

Student Book
pages 190–191

Key points

- Nuclear radiation causes ionisation.
- Common forms of nuclear radiation are alpha, beta and gamma.
- Film badges can monitor the exposure of a person to radiation.

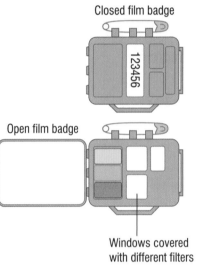

Closed film badge

123456

Open film badge

Windows covered with different filters

The film badge monitors exposure to different types of radiation.

Key words: radioactive, alpha particle, beta particle, gamma radiation, film badge

11.7 What is radioactivity?

- Nuclear radiation comes from the **nucleus** of **radioactive** atoms. It ionises atoms leaving them as charged ions. Ions are atoms that have gained or lost electrons. Forms of nuclear radiation are **alpha** or **beta particles** and **gamma radiation**.

▶ **1** *What is the difference between an ion and an atom?*

Properties of the different types of radiation

	Alpha particle	Beta particle	Gamma ray
What is it?	Two protons and two neutrons	An electron	Very high-energy, electromagnetic wave
Mass compared with a proton	4	$\dfrac{1}{2000}$	0
Charge compared with a proton	+2	−1	0
How ionising is it?	Very ionising	Weakly ionising	Very weakly ionising
Penetrating power	Absorbed by skin, thick paper or 10 cm of air	Passes through skin, thin aluminium or a few metres of air. Absorbed by lead.	Very penetrating. Partly absorbed by dense materials like lead and concrete.

- If molecules in living cells are damaged, the cells may die, turn cancerous or stop working properly. **Film badges** monitor a person's exposure to radiation. They are used by people working with radiation to give information to help limit their exposure.

▶ **2** *Which type of radiation a) has no mass, b) has a positive charge, c) is absorbed by a few metres of air?*

Student Book
pages 192–193

Key points

- Radiotherapy is used to treat cancer.
- Gamma cameras and tracers can diagnose cancer.
- Radiotherapy is not suitable for treating all cancers.
- Patients and doctors should consider the ethical issues before treating cancer with ionising radiation.

11.8 Uses of ionising radiation

- **Radiotherapy** is when cancer cells are killed using ionising radiation. It is only used for some types of cancer, as the cancer cells must be killed but healthy cells left unharmed.
- **Tracers** are radioactive substances eaten, drunk or injected into a patient. A gamma camera detects where the radioactive material collects, which can show obstructions or tumours.
- Cancer treatments kill living cells, and can make patients ill in the short term. Doctors and patients must decide if the benefits of the treatment are greater than the harm caused by the therapy. Treatment does not always cure cancer, but may extend the patient's life or help them feel better.

▶ **1** *What are tracers?*

Key words: radiotherapy, tracer

1 What is the role of platelets in the body?

2 Name an example of a disease caused by:

a bacteria

b a virus.

3 Name three ways you could prevent yourself catching an infectious disease.

4 What do lymphocytes do?

5 State an advantage and possible disadvantage of immunisation.

6 How does an immunisation offer protection against a disease?

7 Describe how X-rays produce an image of broken bones.

8 Which part of the atom does ionising radiation come from?

9 What is an alpha particle made up from?

10 Which types of ionising radiation can penetrate into the body?

11 What equipment is used to detect where a tracer is in the body?

Chapter checklist ✓✓✓

Tick when you have:				Checklist items			
reviewed it after your lesson	☑	☐	☐	Harmful microorganisms	☐	☐	☐
revised once – some questions right	☑	☑	☐	How are diseases spread?	☐	☐	☐
revised twice – all questions right	☑	☑	☑	Body defence mechanisms	☐	☐	☐
Move on to another topic when you have all three ticks				Immunisation	☐	☐	☐
				Vaccination issues	☐	☐	☐
				Medical uses of X-rays	☐	☐	☐
				What is radioactivity?	☐	☐	☐
				Uses of ionising radiation	☐	☐	☐

Student Book
pages 198–199

Key points

- Electrolysis is the breakdown of a compound by electricity.
- Electrolysis involves passing an electric current through a liquid containing freely moving ions (which is called an electrolyte).
- Positive cations are attracted to the negative cathode.
- Negative anions are attracted to the positive anode.

12.1 Electrolysis

- **Electrolysis** is a process that scientists can use to help cause chemical changes. Using electricity, compounds can be broken down and split up. The important parts are the **electrodes**, the power source and the **electrolyte**.
- The electrolyte is often a solution of the substance being broken down. It contains positive **ions** (called **cations**) and negative ions (called **anions**).
- The electrodes are conductors that dip into the electrolyte. The **cathode** is negative and the **anode** is positive.
- The power source is dc (one-way direct current).
- Opposite charges attract, so the cations move toward the cathode (−) and the anions move toward the anode (+). At the cathode, metal cations gain electrons and turn back into metal atoms.

⟫ **1** *Why do anions move toward the anode?*

Key words: electrolysis, electrode, electrolyte

Student Book
pages 200–201

Key points

- Electrolysis can be used to plate metals with other metals in a process called electroplating.
- In electroplating, the cathode is plated with metal from the anode.
- We can write chemical equations to describe what happens at the electrodes during electroplating [H].

12.2 Electroplating

- Electrolysis can be used to **electroplate** (cover) metals with thin layers of other metals. This can make them last longer or look more attractive.
- The object being electroplated is used as a cathode. A pure piece of the metal doing the plating is used as the anode. The electrolyte contains ions of the metal being plated.

⟫ **1** *If you were electroplating an earring, should the earring be the anode or the cathode?*

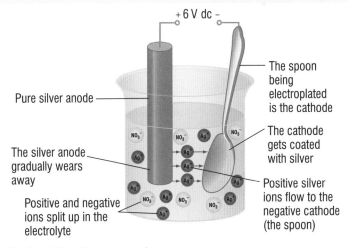

+ 6 V dc −

Pure silver anode

The silver anode gradually wears away

Positive and negative ions split up in the electrolyte

The spoon being electroplated is the cathode

The cathode gets coated with silver

Positive silver ions flow to the negative cathode (the spoon)

Electroplating silver onto an iron spoon

AQA *Examiner's tip*

Remember that the type of ion affects the number of electrons lost or gained, e.g. if a metal forms 2+ ions, then its ions receive 2 electrons from the cathode. The anode is worn away and its atoms lose 2 electrons at the anode. [H]

Electrode equations: Plating silver onto a spoon

At the anode, silver atoms from the anode lose electrons and become cations:
$Ag \longrightarrow Ag^+ + e^-$

At the cathode, silver ions from the electrolyte gain electrons and become atoms of silver metal: $Ag^+ + e^- \longrightarrow Ag$

Higher

Key words: electroplate

Student Book
pages 202–203

12.3 Reasons for electroplating objects

Key points

- Objects are electroplated to make them resistant to corrosion, prevent allergic reactions and make them more attractive.
- The chemicals used in electroplating are hazardous. Many safety procedures are in place to reduce the risk of harm.

- Here is a summary of reasons why materials are electroplated:

Reason for electroplating	Examples
Protection from **corrosion** (**rusting**) – plating with a less reactive metal	**Chromium** is used to chrome plate parts of cars and motorcycles. Tin is used to plate the insides of steel food cans
Reduce the effects of **allergies**	Many people are allergic to cheap **nickel** jewellery. Plating the jewellery with gold or silver prevents an allergic reaction
Decoration	Silver plating knives and forks. This also helps them resist corrosion

- Often, the electrolytes used in electroplating are toxic or corrosive. **Protective clothing** needs to be worn at all times during the process to prevent **chemical burns**, poisoning, or respiratory problems.

➡ 1 *Why are metal kettles often plated with chromium?*

Key words: allergy, chemical burn

Student Book
pages 204–205

12.4 Developing smart materials

Key points

- Smart materials bring a wide variety of new properties to products. They can change depending on their environment.
- Smart materials include memory metals, self-healing paints, thermochromic materials and photochromic materials.

Bump up your grade

Explaining why a smart material is used will get you good marks. To secure even more marks, you need to be able to compare it to existing materials when you explain why it is used.

Key words: memory metal, self-healing paint, thermochromic pigment, photochromic pigment

- Smart materials can change in response to their surroundings. They are used in medicine, dentistry, cars, jet engines, robots, packaging, and around the home. Here are some examples:

Type of material	What it does	Examples
Memory metals	Made to 'remember' different shapes. Different temperatures make them switch between shapes	Unbreakable spectacle frames, self-tightening dental braces, stents (wire tubes that hold blood vessels open), 'muscle wires' in robots
Self-healing paint	Surfaces that can repair themselves when scratched. Heat or UV light triggers the process	Used in some car paints and phones. It isn't widely used yet and is still quite expensive
Thermochromic pigments	Inks that change colour at different temperatures	Smart packaging and safety labels tell us if food or medicines were stored at the wrong temperature. Baby products that warn us if something is too hot
Photochromic pigments	Inks that change colour when exposed to light	UV sensor bracelets to prevent sunburn. Rear-view mirrors in cars that darken so the driver isn't dazzled by headlights. Spectacles that darken in brighter light

➡ 1 *Describe two medical uses for smart materials.*
➡ 2 *Suggest why self-healing paint isn't used to paint the walls of houses.*

Student Book
pages 206–207

Key points

- Superconductors have almost zero electrical resistance at very low temperatures.
- They can make very strong electromagnets.
- They are used in MRI scanners and Maglev trains.

AQA **Examiner's tip**

Make sure you are able to discuss the advantages and disadvantages of different materials.

12.5 Superconductors

- **Resistance** measures how hard it is for an electric current to flow. **Superconductors** are materials with almost zero resistance to electricity at very low temperatures. When the power supply is turned off, the current keeps flowing. This reduces energy wastage in the circuit.

▸ **1** *What is special about superconductors?*

Superconductors are used to make very strong **electromagnets**. These are used for:

- **MRI scanners**, which are used to make very detailed scans inside a person's body
- **Maglev trains**, which are trains that can go extremely fast as they float just above the rail tracks.

Key words: resistance, superconductor, electromagnet, MRI scanner

Student Book
pages 208–209

Key points

- New and exciting materials are being developed by research scientists for new technologies or to replace traditional materials.
- Many modern materials are often more expensive and more specialised than traditional materials.
- Producing and disposing of modern materials may affect the environment more than traditional materials.

12.6 Advantages and disadvantages of modern products

- Scientists have developed modern materials with special properties for specific uses. They are more expensive than traditional materials, but are more suited for their purpose.
- Self-healing paints save money on repair bills. Smart paints are more expensive than traditional paints, and must be applied by specialists.
- Superconductors can make smaller, stronger electromagnets than traditional materials. They can only be used at extremely cold temperatures.
- **Smart materials** automatically change if their surroundings change, and are useful for medical applications. They are more expensive than traditional materials.
- **Chromic materials** automatically provide information about the conditions they are exposed to. They are toxic and hard to dispose of. Using chromic materials affects the environment more than using traditional products.

▸ **1** *Write down one advantage of chromic materials.*

Key words: smart material, chromic material

1 Add names to these descriptions:

   a ionic substance broken down in electrolysis

   b a negative electrode

   c attracted to a positive electrode

   d coating metals with other metals using electrolysis.

2 You are electroplating a cheap nickel statue with gold. What would you use for:

   a the anode   b the cathode   c the electrolyte?

3 Why do ions change into metal atoms at the cathode?   [H]

4 Why do metal atoms change into ions at the anode?   [H]

5 Describe three uses for photochromic pigments.

6 How can smart packaging tell us if medicines are not safe to use?

7 How does the resistance of superconductors change with temperature?

8 Write down two uses of superconductors.

9 Explain why smart materials are suitable for medical uses.

10 Explain how self-healing paints help save money on repair bills.

Chapter checklist	✓ ✓ ✓
Tick when you have:	
reviewed it after your lesson ☑ ☐ ☐	Electrolysis ☐ ☐ ☐
revised once – some questions right ☑ ☑ ☐	Electroplating ☐ ☐ ☐
revised twice – all questions right ☑ ☑ ☑	Reasons for electroplating objects ☐ ☐ ☐
Move on to another topic when you have all three ticks	Developing smart materials ☐ ☐ ☐
	Superconductors ☐ ☐ ☐
	Advantages and disadvantages of modern products ☐ ☐ ☐

**Student Book
pages 212–213**

Key points

- Selective breeding means choosing the best organisms to breed, producing offspring with desired characteristics.

13.1 Selective breeding

- Farmers select their best animals or plants to breed, to produce the highest yields. This is called **selective breeding**. For example, they select dairy cattle that produce lots of milk or tomato plants that produce a high yield.

The farmer selects the ewe with the longest fleece; he breeds this with his best long-fleeced ram ➡ The farmer chooses the best ewe and breeds again with his best ram ➡ Eventually all the sheep have the desired characteristic of a good-quality, long fleece

> **1** *What does the term 'selective breeding' mean?*

- Selective breeding reduces the number of genes (the gene pool) from which a particular strain of species is created. If a new disease arises, an organism in that gene pool may not contain the gene for resistance to this disease. This could result in extinction.
- Selective breeding reduces variation. For example, pedigree dogs are selectively bred. They have the desired characteristics of their breed, but many suffer from health problems because of this.

> **2** *Name a disadvantage of selective breeding.*

Key words: selective breeding

**Student Book
pages 214–215**

Key points

- Genetic engineering changes an organism's genes to produce desired characteristics.
- Genetically modified plants include frost-resistant tomatoes and insect-resistant Bt corn.

AQA *Examiner's tip*

Remember – genetic engineering is also known as genetic modification. Foods that are manufactured from genetically engineered crops are often referred to as 'GM foods'.

Key words: genetic engineering, genetic modification

13.2 Genetically modified food

- Selective breeding produces animals with desired characteristics. It is a slow process with unpredictable outcomes.
- Scientists can alter the combinations of genes in an organism to produce the desired characteristics. This is called **genetic engineering** (or **genetic modification**). It can happen in one generation.
- Genetic modification works by putting foreign genes into plant or animal cells at an early stage in their development. As the organism develops, it displays the characteristics of the foreign genes. For example, *Bacillus thuringiensis* contains a gene which codes for a toxin that kills insects. This can be inserted into maize, creating 'Bt corn'. Insecticides are no longer needed.

> **1** *Name an example of a genetically engineered plant.*

A useful gene is removed from the nucleus of a donor cell.
The foreign gene is then put into a circular piece of DNA called a **plasmid**. This is now known as a piece of **recombinant DNA**.
The recombinant DNA is put into a bacterial cell.
Plant cells are infected with the bacteria. The foreign gene becomes integrated with the DNA of the plant cells.
The plants cells are placed in a growing medium to grow into plants. These plants will have the desired characteristics.

How crops are genetically modified

Student Book
pages 216–217

13.3 Genetic engineering and cloning

Key points

- Genes can be inserted into bacteria to make them produce useful chemicals, including insulin and vaccines.
- A clone is a genetically identical copy of an organism.
- Genetic engineering alters the genes of an organism; cloning produces exact copies of an organism.

- When microorganisms reproduce, they produce genetically identical copies of themselves – **clones**. Bacteria have been genetically engineered to produce chemicals such as vaccines, antibiotics and the hormone **insulin**. Bacteria are made to produce insulin in the following way:

  1 Small circles of DNA called **plasmids** are found in bacteria. These are modified to include a piece of human DNA which contains instructions for making insulin.

  2 The plasmid is then inserted into the bacteria *Escherichia coli*. The bacteria now produce insulin.

  3 The bacteria multiply quickly, and produce large quantities of insulin in a fermenter.

  4 Bacteria are then killed by heat (sterilisation), leaving behind the insulin. The insulin is used to control blood glucose levels of people suffering from diabetes.

➡ 1 *What is a plasmid?*

- Micropropagation (plant tissue culture) allows the rapid production of many genetically identical plants (plant clones). Very little space is needed, and the plants are disease-free as they are grown in controlled environments. However the gene pool is reduced, increasing the risk of disease destroying a species.

➡ 2 *Name two advantages of micropropagation.*

- Human cells can be cloned in the laboratory and used for research into diseases and new drugs. Tissue culture can be used to produce cartilage and skin for grafting.
- Small pieces of a person's tissue are collected and transferred to an artificial, sterile environment. The cells continue to survive and function. When the skin tissue has grown to the desired size, it can be used to repair damage.
- As the tissue is genetically identical to the donor material, the body will not recognise it as being 'foreign'. This minimises the risk of rejection.

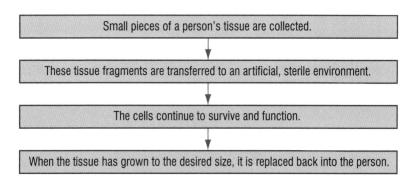

Small pieces of a person's tissue are collected.

↓

These tissue fragments are transferred to an artificial, sterile environment.

↓

The cells continue to survive and function.

↓

When the tissue has grown to the desired size, it is replaced back into the person.

➡ 3 *Name an example of human cloning that is used regularly in medical treatments.*

Key words: clone, insulin, plasmid

1. A tissue sample is scraped from the parent plant (only a few cells are needed).

2. Tissue samples are placed on an agar plate containing nutrients and auxins. Auxins are plant growth hormones.

3. Samples grow into tiny plants (plantlets).

4. Plantlets are planted in compost, and grown in a greenhouse where they develop into full-sized plants.

The process of micropropagation

Student Book
pages 218–219

13.4 Ethics of genetic engineering and cloning

Key points

- Genetic research is highly controversial, as the techniques involve altering the genetic material present in an organism.

- A 'designer baby' develops from an embryo that has been created specifically for a desired characteristic.

- In gene replacement therapy, a faulty gene will be replaced with a healthy gene. This technique has the potential to cure many genetically inherited disorders.

- There is a lot of controversy surrounding genetic research. Supporters say that scientists have the potential to tackle huge issues. These include feeding the world's increasing population, and providing cures and treatments for a number of diseases. Others believe these techniques are unethical, and should be banned.

Arguments *for* genetic research:	Arguments *against* genetic research:
Plants and animals can be produced with desired characteristics such as higher yields, disease and frost resistance	May reduce an animal's life span or affect an animal's quality of life.
Seedless fruits produced – desired by consumers	If seedless species interact with wild plants it may result in wild plants not producing seeds
Scientists could clone more individuals of a species threatened with extinction	It is not a natural process
May offer cures for genetically inherited disorders	Techniques could be used to change a person's appearance or characteristics
Medicines can be produced cheaply and quickly using bacteria	
People would not reject cloned copies of their own organs, which could be used to replace their damaged or diseased organs	

- During *in vitro* fertilisation (IVF), eggs are fertilised with sperm in the laboratory. The embryos are then implanted into the woman's uterus. This technique allows some couples with fertility problems to have children. If a couple is at risk of an inherited disease, the embryos created can be screened for genetic disorders. Only healthy embryos are selected to be implanted.

- People are concerned that couples may want to choose 'designer babies' with desirable qualities – such as choosing to have a girl, if the couple already have boys.

⟫ **1** *Name an advantage and a disadvantage of IVF.*

- Genetically inherited disorders, such as cystic fibrosis, are impossible to cure using traditional medical techniques. **Gene replacement therapy** is the replacement of the faulty gene (which results in a disorder) with a healthy gene in a person's cells. This would lead to the disease being cured.

- Some people are concerned that gene replacement therapy may be used for non-medical purposes. For example, it could be used to improve your appearance, or to make you stronger or faster.

⟫ **2** *Name an advantage and a disadvantage of gene replacement therapy.*

Key words: *in vitro* fertilisation (IVF), gene replacement therapy

AQA Examiner's tip

An exam question may ask you to 'discuss' an aspect of genetic engineering. To successfully answer this type of question, you should present arguments for and against the technique, and be able to justify your answer using scientific reasons. A range of answers will be acceptable – there is no right or wrong answer.

1 When a bacterium replicates, a clone is produced. What does this mean?

2 For each of the following organisms, describe the characteristics that a farmer would choose to selectively breed for the highest yields:

    a dairy cattle

    b beef cattle

    c chickens

3 Name an example of an animal tissue that is cloned for medical purposes.

4 What is meant by 'genetic engineering'? Why is this advantageous to selective breeding?

5 What is meant by the term 'designer baby'?

6 What is meant by 'gene replacement therapy'?

7 Name two arguments for, and two arguments against, genetic engineering.

8 Describe how plants can be cloned by the process of micropropagation.

9 Produce a flow diagram to show how cattle can be selectively bred to produce higher milk yields.

10 How are bacteria genetically engineered to produce insulin?

Chapter checklist	✓	✓	✓
Tick when you have:			
reviewed it after your lesson	✓	☐	☐
revised once – some questions right	✓	✓	☐
revised twice – all questions right	✓	✓	✓
Move on to another topic when you have all three ticks			

	✓	✓	✓
Selective breeding	☐	☐	☐
Genetically modified food	☐	☐	☐
Genetic engineering and cloning	☐	☐	☐
Ethics of genetic engineering and cloning	☐	☐	☐

**Student Book
pages 224–225**

14.1 Producing greenhouse gases

Key points

- Greenhouse gases include carbon dioxide, methane and nitrous oxide.
- Carbon dioxide is produced when fuels are burned.
- Methane is mainly produced when bacteria break down organic matter in the absence of oxygen.
- Nitrous oxide is produced by nitrogen-based fertilisers from intensive farming and by car engines.

- **Carbon dioxide**, **methane** and **nitrous oxide** are examples of **greenhouse gases**. Greenhouse gas production has increased due to electricity generation, increased travel and intensive farming methods.

Greenhouse gas	How it is produced
Carbon dioxide	When fuels are burned, e.g. in motor vehicle engines and power stations generating electricity
Methane	When organic materials rot in the absence of oxygen, e.g. in landfill sites, rice paddy fields, and other forms of agriculture
	From coal mining activities
Nitrous oxide	During intensive farming because large amounts of fertiliser or animal waste cause **denitrification** (when bacteria convert nitrates in the soil to nitrous oxide)
	When fuels burn at very high temperatures in car engines

AQA *Examiner's tip*

Make sure you can match each greenhouse gas with the way it is produced.

▐▐▐➡ 1 *Which greenhouse gases are produced*
    a) when fuels are burned   b) due to farming?

Key words: carbon dioxide, methane, nitrous oxide, denitrification

**Student Book
pages 226–227**

14.2 The effects of greenhouse gases

Key points

- The greenhouse effect is when energy is retained in the atmosphere by greenhouse gases.
- Global warming is the increase in the average temperature of our atmosphere, causing climate change.
- Many countries in the world agreed to cut their emissions of greenhouse gases as a result of the Kyoto Protocol.

- Greenhouse gases in the atmosphere absorb long-wavelength (infrared) radiation from the Earth's surface, keeping the energy in the atmosphere. This is the **greenhouse effect**.
- The radiation is re-radiated, which increases the average atmospheric temperature.
- **Global warming** occurs when more radiation is absorbed than is emitted from the atmosphere. Global warming changes the climate.

How global warming is caused

- The Kyoto Protocol is an international legal agreement. Most countries in the world agreed to reduce the amount of greenhouse gases they produce. The aim is to stabilise global warming and prevent **climate change**. Australia and the USA refused to sign.

▐▐▐➡ 1 *What is the Kyoto Protocol?*

Key words: greenhouse effect, global warming

**Student Book
pages 228–229**

14.3 Threats to the countryside

- Artificial **fertilisers**, **pesticides** and **herbicides** are used in **intensive farming**.

Type of chemical	Effect on crops
Fertiliser	Provides nutrients
Pesticide	Kills insects that may eat the crop
Herbicide	Kills weeds that may compete with the crop

▐▐▶ **1** *Why do farmers use pesticides?*

- If fertilisers **leach** into lakes and rivers, algae grow rapidly. This reduces oxygen levels in the water, which can kill plants and animals living in it. This is **eutrophication**. Look at the diagram below.

Key points

- Fertilisers, pesticides and herbicides are used to increase crop production in intensive farming.

- If fertiliser drains into rivers and lakes it can cause eutrophication.

- Indicator species can be used to monitor the presence of pollution. Their numbers increase or decrease in the presence of certain pollutants.

Fertiliser run-off

Excess fertiliser dissolves in water and runs off the land into waterways (leaching).

Algae in waterways grow rapidly because of the fertilisers. They cover the water surface.

Plants beneath the surface die and are decomposed by microorganisms, which use up oxygen in the water. Fish and animals in the water die.

The main steps in eutrophication

- **Indicator species** are very sensitive to changes in the pollution levels. Monitoring changes in the levels of indicator species keeps track of changes in the pollution levels of ecosystems.

Indicator species	Type of pollution detected	How changes monitor pollution levels
Bloodworm	Water	High levels indicate high pollution levels
Water louse	Water	High levels indicate high pollution levels
Sludgeworm	Water	High levels indicate very high pollution levels and very low oxygen levels
Rat-tailed maggot	Water	High levels indicate very high pollution levels
Lichen	Air	Low levels of lichen indicate high levels of sulfur dioxide or ozone

▐▐▶ **2** *How do indicator species help scientists monitor pollution levels?*

Key words: fertiliser, pesticide, herbicide, intensive farming, leach, eutrophication, indicator species

AQA *Examiner's tip*

Make sure you can explain the stages occurring during eutrophication. In an exam you may be asked to describe all the steps involved. Draw a flow chart to help you remember.

Student Book
pages 230–231

14.4 Disposing of our waste

Key points

- Disposing of products in landfills and incinerators damages the environment.
- Polymers like EVOH and PVOH are used in packaging because they are water soluble and will biodegrade. [H]
- Using biopolymers, photo-degradable materials and oxo-degradable materials speeds up biodegradation. [H]

Landfills and incinerators

- Most waste in the UK goes to landfill sites. These are cheap to set up as they are just deep holes in the ground. They can release gases and liquids that need to be carefully removed. They also take up a lot of space.
- **Incinerators** get rid of waste by burning it. They take up less space than landfills, but the waste gases are often toxic. Sometimes, the heat released in incinerators can be used to generate electricity.

1 *Describe two advantages of incinerators over landfill sites.*

Degrading waste

- Some waste, like food and paper, can be broken down by microbes in the environment (**biodegrade**). **Non-biodegradable** polymers take up a lot of the space in landfill sites.

New polymers can be used to accelerate breakdown

Higher

Name of polymer	Why it degrades faster
Polyvinyl alcohol (PVOH) and **ethylene vinyl alcohol (EVOH)**	Water soluble, so break down in the environment. (This also shortens their shelf-life.)
Polylactic acid (PLA)	PLA is made from potato starch, so can be digested by microbes.
Oxo-degradable polymers	Contain an additive that breaks down in air and helps microbes start digesting them.
Photo-degradable polymers	Break down when exposed to light.

2 *What takes up most of the space in landfill sites?*

3 *Suggest the monomer used to make polylactic acid (PLA).*

Key words: biodegrade, polyvinyl alcohol (PVOH), ethylene vinyl alcohol (EVOH), polylactic acid (PLA), oxo-degradable polymer, photo-degradable polymer

AQA *Examiner's tip*

If you are asked to compare landfills and incinerators, remember they both have advantages and disadvantages.

1 Describe three reasons why greenhouse gas emissions have increased.

2 Explain why denitrification increases the levels of nitrous oxides.

3 What is the difference between the greenhouse effect and global warming?

4 Explain why increases in greenhouse gases affect the climate.

5 Why was it important that many of the countries producing greenhouse gas emissions signed the Kyoto Protocol?

6 Why do you think some countries refused to sign the Kyoto Protocol?

7 State one advantage and one disadvantage of farmers using artificial fertilisers.

8 How could an indicator species show whether there is sulfur dioxide pollution in an area?

9 Why are most polymers non-biodegradable?

10 Describe how biomass solves the problem of non-biodegradable carrier bags.

[H]

11 Describe three ways to help polymers degrade.

[H]

Chapter checklist	✓ ✓ ✓

Tick when you have:				Producing greenhouse gases	☐	☐	☐
reviewed it after your lesson	✓	☐	☐	The effects of greenhouse gases	☐	☐	☐
revised once – some questions right	✓	✓	☐				
revised twice – all questions right	✓	✓	✓	Threats to the countryside	☐	☐	☐
Move on to another topic when you have all three ticks				Disposing of our waste	☐	☐	☐

Student Book
pages 234–235

15.1 Conduction

- Heat energy always travels from a hotter to a cooler place. Heat energy is transferred in solids by **conduction**. Particles in fixed positions vibrate more when heated, knocking into neighbouring atoms, and passing on energy.
- Conduction takes place more quickly if there is a large temperature difference, if the object is made from a good heat conductor, or if the object has a large cross-section.

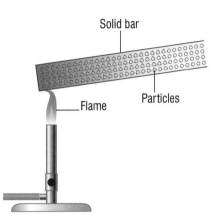

Solid bar

Particles in a solid vibrate in fixed positions. They vibrate more when the solid heats up, passing energy to their neighbours.

Flame

Particles

- Metals are **good heat conductors**. Poor heat conductors (**insulators**) are non-metals like wood, fibreglass, wool, and materials containing air pockets.
- Heat losses in the home are reduced using double glazing, fibreglass insulation in lofts, jackets fitted around hot water tanks, cavity wall insulation.

▌▌▶ **1** *Why are carpets good insulators?*

Key words: conduction, insulator

Key points

- Conduction is when heat is transferred between neighbouring particles.
- Metals are good heat conductors.
- Non-metals and gases are poor heat conductors, or good insulators.

Bump up your grade

Make sure you can suggest suitable materials to alter the rate of conduction and control how fast objects heat up or cool down.

AQA **Examiner's tip**

Make sure you know examples of insulators and can describe how they are used in the home.

Student Book
pages 236–237

15.2 Convection

- When part of a **fluid** is heated, it expands, becoming less dense. Warmer fluids rise to float above cooler fluids. **Convection currents** are set up if the fluid is heated from the base or cooled from the top.
- Unwanted transfers of heat energy in the home are reduced by using **draught excluders**, keeping doors and windows closed, blocking unused fireplaces.

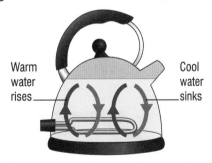

Warm water rises

Cool water sinks

Convection currents spread heat evenly through water

▌▌▶ **1** *Explain why draught excluders reduce the transfer of heat energy.*

Key words: fluid, convection current

Key points

- Convection occurs in liquids and gases (fluids) when particles move, carrying heat with them.
- Convection currents spread heat in fluids when they are heated from the base or cooled from the top.

AQA **Examiner's tip**

If the question asks about convection currents in liquids, make sure your answer refers to liquids (and not gases).

15.3 Radiation

Student Book
pages 238–239

Key points

- Radiation is emitted when infrared waves transfer heat energy from the surface of objects.
- Hot objects radiate energy more quickly than cooler objects.
- Black matt objects radiate more energy in a given time than white shiny objects.

AQA *Examiner's tip*

A black surface helps a cool object heat up quickly, and also allows a hot object to cool down quickly.

- Heat energy is transferred in waves from the surface of hot objects by **infrared radiation**. Objects cool down if they **radiate** heat, and warm up if they **absorb** radiation.

Heat energy is lost by radiation more quickly if:	Heat energy is lost by radiation more slowly if:
The object is much hotter than the surroundings	The object is only slightly hotter than the surroundings
The surface colour is matt black	The surface colour is shiny white
The surface area is large	The surface area is small

- Heat energy losses in the home are reduced by putting foil on walls behind radiators.

▐▋▋➤ **1** *Explain why the insides of ovens are painted black.*

Key words: radiate, absorb

15.4 Will you save money?

Student Book
pages 240–241

Key points

- Payback time = cost of installation ÷ annual savings
- Efficient equipment is not always cost-effective to install.
- Most methods to reduce heat losses in buildings increase the insulation installed.

AQA *Examiner's tip*

Be careful when changing fractions of a year into months.

- Householders can save costs by reducing energy losses, e.g. by:
  - reducing heat losses
  - using efficient equipment
  - using equipment more efficiently.
- **Payback time** is the time for the cost of installing energy saving measures to match the savings they generate. It is calculated using:

  payback time = cost of installation ÷ annual savings
- It is not always worth installing the energy saving measure. For example, double glazing is expensive to install and does not reduce heat losses much, so its payback period is very long.

▐▋▋➤ **1** *What is the payback time if draught excluders cost £10 to install and save the homeowner £30 per year?*

- Some boilers are very **efficient**, wasting very little energy. However, the new boiler is only **cost-effective** if it costs less to install than the savings it generates.

Bump up your grade

To get good marks you should be able to describe some energy saving measures. To get even better marks, you will choose the best one and explain in detail why you chose it.

Key words: payback time, efficient, cost-effective

**Student Book
pages 242–243**

15.5 U-values

Key points

- U-values measure the rate of heat loss through a material.
- Low U-values mean heat is transferred more slowly through the material.
- Most buildings use materials with low U-values.

- Different building materials transfer heat more quickly than others.
- **U-values** compare the rate of heat loss through different materials:
  - good heat conductors like metals have a high U-value because they transfer heat quickly
  - insulators have low U-values.
- Architects aim to specify materials with low U-values to reduce heat losses in winter and so keep a home warm.

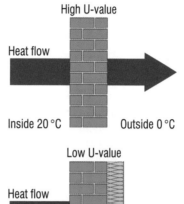

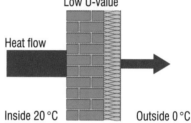

Heat flows more slowly if the U-value is low

In the home	Good choice of material: low U-value	Bad choice of material: high U-value
Window frame	Wood or PVC	Aluminium
Wall	Cavity wall (two layers of brick separated by an air gap)	Glass
Window	Double glazing	Single glazing
Roof	Thatch	Lead tiles

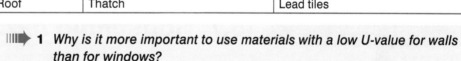

1 *Why is it more important to use materials with a low U-value for walls than for windows?*

Key words: U-value

Student Book
pages 244–245

15.6 Pollution in the home

- Central heating and air conditioning mean that some offices and houses don't open their windows very often. This can lead to a build up of air pollutants that cause respiratory illnesses.

Key points

- Closed air systems in homes can lead to a build up of indoor pollutants.
- Indoor air pollutants include dust, mould and spores, pollen, smoke, fumes from household products and carbon monoxide.
- Some of the symptoms of exposure to high levels of indoor pollution are asthma, headaches, tiredness, nausea, itchy nose and sore throat.

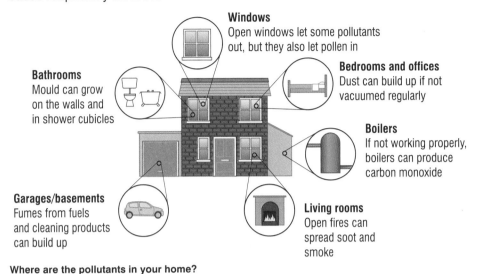

Windows
Open windows let some pollutants out, but they also let pollen in

Bathrooms
Mould can grow on the walls and in shower cubicles

Bedrooms and offices
Dust can build up if not vacuumed regularly

Boilers
If not working properly, boilers can produce carbon monoxide

Garages/basements
Fumes from fuels and cleaning products can build up

Living rooms
Open fires can spread soot and smoke

Where are the pollutants in your home?

Bump up your grade

To get good marks, you must be able to describe the effects of indoor pollutants as well as naming them.

- Different pollutants can affect us in different ways:

Pollutant	Its effects
Mould, **spores**, **pollen** and dust	Allergic reactions, itchy nose
Soot and smoke	Lung damage, sore throat, watery eyes
Fuels and cleaning products	Nausea, tiredness, headaches
Carbon monoxide	Tiredness, death

▶ **1** *Which parts of a home might you find mould and spores in?*

Key words: mould, spore, pollen

Student Book
pages 246–247

15.7 Household hazards

- Many chemicals we use every day are harmful unless used carefully. Hazard symbols are printed on products to help us understand potential risks. These are the most common symbols you will see:

Key points

- Many household products contain hazardous chemicals.
- Products must be clearly labelled with hazard symbols.
- Exposure to hazardous chemicals can cause dizziness, rashes, headaches and other symptoms.

Harmful	Irritant	Corrosive
Can damage your health. Don't inhale, eat or drink them, or allow them to touch bare skin.	Can cause rashes. Take the same precautions as harmful substances.	Causes chemical burns. Use gloves and eye protection.

Environmental hazard	Flammable	Toxic
Dispose of this carefully as it can kill plants and animals.	Keep sealed and away from heat sources.	Poisonous. Use gloves, don't drink this.

⫸ 1 *What might happen to your skin if you spilled corrosive oven cleaner on it?*

⫸ 2 *How do we know the risks of using household chemicals?*

**Student Book
pages 248–249**

15.8 The silent killer

Key points

- Incomplete combustion happens when a fuel is burned without enough oxygen.
- Incomplete combustion produces carbon monoxide and carbon (soot).
- Carbon monoxide is poisonous because it replaces the oxygen in your red blood cells.

Incomplete combustion

- The fuels we use in homes need a good supply of oxygen to burn properly. Without enough oxygen, **incomplete combustion** takes place:

hydrocarbon fuel + oxygen ⟶ carbon dioxide + water + carbon monoxide + carbon

- Incomplete combustion releases less energy than complete combustion. It's also very dangerous, as carbon monoxide gas is poisonous.

⫸ 1 *What effect would incomplete combustion have on the efficiency of a boiler?*

Carbon monoxide poisoning

- Carbon monoxide is a colourless and odourless gas. It binds to **haemoglobin** in your **red blood cells**. It does this 200 times more strongly than oxygen, so it can quickly replace a lot of the oxygen in your body. The first symptoms are tiredness and headache, followed by coma and death.

⫸ 2 *Suggest why carbon monoxide poisoning is sometimes mistaken for a cold.*

Preventing carbon monoxide poisoning

- The carbon produced in incomplete combustion builds up as **soot**. This makes it easier to see if incomplete combustion is going on. Regularly servicing gas and oil boilers helps ensure they are properly ventilated and performing complete combustion.

Key words: haemoglobin

AQA Examiner's tip

Incomplete combustion is burning without *enough* oxygen. You will lose marks for saying *no* oxygen. (With no oxygen a fuel cannot burn at all.)

Student Book
pages 250–251

15.9 Radon gas

Key points

- Radon is a radioactive gas that can cause cancer.
- Radon seeps from rocks under buildings that contain uranium and plutonium.
- Radon levels can be reduced by ventilating the building.

- **Radon** gas is a radioactive gas that seeps naturally out of rocks like granite. These rocks contain large concentrations of **radium** or **uranium**. In homes built on these rocks, the amount of radon gas can build up and become a pollutant.
- The radioactivity in radon gas can damage cells in lungs, which become cancerous. Since smokers have already damaged the cells in their lungs, they are affected much more by lung cancer caused by radon gas than non-smokers.
- In high-risk areas, people can reduce the amount of radon gas in homes by improving the ventilation or sealing floors. Some houses have an extractor fan fitted in their basement.

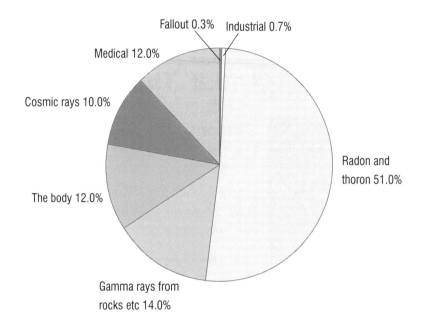

Fallout 0.3%   Industrial 0.7%
Medical 12.0%
Cosmic rays 10.0%
The body 12.0%
Radon and thoron 51.0%
Gamma rays from rocks etc 14.0%

The main sources of radioactivity we are exposed to every day

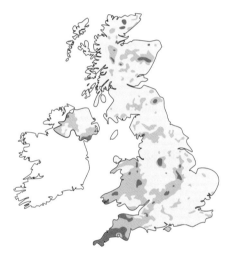

Radon occurs in different amounts across the UK. The darker the colour on the map, the higher the level of radon.

Based on information supplied by The Department for Environment, Food and Rural Affairs.

⊪➡ 1 *Why isn't radon gas a pollutant in all parts of the country?*

AQA *Examiner's tip*

Make sure you can explain why radon gas is a health hazard.

Key words: radon, radium, uranium

1 Name three types of heat energy transfer.

2 Explain how convection currents are set up in liquids.

3 Why doesn't convection take place in solids?

4 Explain three ways that light-coloured curtains reduce heat losses.

5 Explain why houses in hot countries are often painted white.

6 A homeowner installs a new boiler. It costs £10 000 and annual savings are £500. What is its payback time?

7 Why is it better to install energy saving measures with short payback times?

8 Why do building materials with low U-values help keep a home cool in summer as well as warm in winter?

9 Smokers living in homes affected by radon gas are more likely to be affected by the radon than non-smokers. Explain why.

10 Describe how living organisms can cause air pollution indoors.

11 Which indoor air pollutants might make you feel nauseous?

12 Compared with complete combustion, what two extra substances are produced during incomplete combustion?

13 Why is carbon monoxide poisonous?

Chapter checklist		✓	✓	✓
Tick when you have:				
reviewed it after your lesson	✓ ☐ ☐			
revised once – some questions right	✓ ✓ ☐			
revised twice – all questions right	✓ ✓ ✓			
Move on to another topic when you have all three ticks				

	✓	✓	✓
Conduction	☐	☐	☐
Convection	☐	☐	☐
Radiation	☐	☐	☐
Will you save money?	☐	☐	☐
U-values	☐	☐	☐
Pollution in the home	☐	☐	☐
Household hazards	☐	☐	☐
The silent killer	☐	☐	☐
Radon gas	☐	☐	☐

1 A farmer wanted to ensure that his chickens always laid lots of eggs.
  a Suggest how he could achieve this using selective breeding.   *(3 marks)*
  b Explain the advantage of genetic modification over selective breeding.   *(2 marks)*

2 Ionising radiation can be helpful but also harmful.

The diagram shows what material stops each type of radiation.

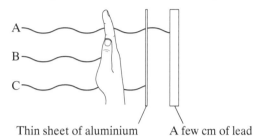

Thin sheet of aluminium   A few cm of lead

  a Name each type of wave A–C.   *(3 marks)*
  b State a use for gamma radiation and describe why it is useful.   *(2 marks)*
  c People who work with ionising radiation wear film badges. State which type of radiation cannot be detected by a film badge and explain why.   *(2 marks)*

3 Doctors can prescribe many medicinal drugs.
  a Why might you need to go to a doctor even after taking a paracetamol tablet?   *(2 marks)*
  b Why would a doctor not prescribe a course of antibiotics for the flu?   *(2 marks)*
  c Why would a doctor want you to take a whole course of antibiotics?   *(2 marks)*
  d Why do doctors not want to prescribe antibiotics too often?   *(2 marks)*

4 Match **two** problems caused by smoking tobacco and **two** problems caused by drinking alcohol.

Tobacco	Long-term damage to brain
	Lung cancer
	Reduces oxygen capacity of the blood
Alcohol	Slows down reactions
	Causes stomach ulcers

   (4 marks)

5 Our body has a defence system to get rid of any pathogens that enter.
  a *In this question you will be assessed on using good English, organising information clearly and using specialist terms where appropriate.*
     How does a vaccination protect against pathogens?   *(6 marks)*
  b Describe the part of the blood that helps to stop pathogens entering the body.   *(2 marks)*

6 Materials scientists are constantly trying to improve products. Match the new product to a possible use.

Product	Use
Chromic material	Aircraft so it doesn't matter if it is scratched
Self-healing paint	Electromagnets
Smart material	Shrink-wrap packaging
Superconductor	Rear-view mirrors

(3 marks)

7 Some metals need to be coated to make them more durable.
  a Describe how metals can be coated using electrolysis. *(4 marks)*
  b **i** Copy and complete this equation to show what happens at the cathode:

$$M^{n+} + ne^- \longrightarrow \;\ldots$$

                                                                   [H] *(1 mark)*
    ii Give the equation to show what happens at the anode. **[H]** *(2 marks)*
  c Read the label from a container of zinc-based paint and answer the question.

> Resistant against corrosion
> Protects even when scratched
> Poisonous

    Give a reason for and against using zinc to coat a bucket for use in a garden. *(2 marks)*

8 The transport industry contributes to global warming.
  a Describe how cars contribute to global warming. *(2 marks)*
  b Describe how greenhouse gases contribute to global warming. *(2 marks)*
  c Name **one** greenhouse gas other than the main gas produced by cars. State **one** of the sources of the named gas. *(2 marks)*

AQA Examiner's tip

There are two other greenhouse gases that you need to know about. See if you can give details about how they are both produced just to test yourself.

Answers

Unit 1, Theme 1

1 Our changing planet and universe

▶ 1.1

1 Advantage of a space-based telescope: there is no interference from the atmosphere. Disadvantages (one of): very expensive to build, hard to maintain, etc.

2 Telescopes can detect radio waves, microwaves, infrared radiation, visible light, ultraviolet radiation, X-rays and gamma rays.

▶ 1.2

1 The red-shift is when light waves are stretched when a light source moves away from an observer.

2 **Two** pieces of evidence are the background microwave radiation and the red-shift from galaxies.

▶ 1.3

1 For example, a star produces and emits light and radiation, a planet does not produce its own light and radiation.

2 It is likely that there is life on other planets as there are so many other solar systems, making it highly probable there are some with planets that could have liquid water on their surface.

▶ 1.4

1 Four: core, mantle, crust, atmosphere

▶ 1.5

1 Heat from nuclear reactions deep inside the Earth

▶ 1.6

1 They used CO_2 in photosynthesis.

▶ 1.7

1 Greenhouse gases in the atmosphere can absorb it.

Answers to end of chapter questions

1 Different objects at different temperatures in the universe emit different frequencies of electromagnetic radiation. Our eyes can only detect visible light.

2 There is no distortion caused by the atmosphere for space-based telescopes.

3 The red-shift, which shows that all galaxies move away from us, and more distant galaxies move away faster.

4 Waves from a moving source are stretched or squashed. Light from a source moving away is stretched and appears redder because red light has a longer wavelength.

5 Nuclear reactions heat the mantle $\longrightarrow$ convection of mantle $\longrightarrow$ plates move around $\longrightarrow$ plates slip under each other and melt $\longrightarrow$ molten rock rises to surface as volcano

6

Name of layer	Location	Description
Crust	Outer layer	Solid rock, cracked into plates
Mantle	Beneath the crust	Slow-flowing rock
Inner core	In the centre of the Earth	Solid nickel and iron
Outer core	Surrounds the inner core	Liquid nickel and iron

7 Because this gas is released by volcanoes and early Earth was very volcanic

8 The oceans formed when water vapour in the atmosphere condensed.

9 Necessary: for photosynthesis and keeping Earth warm. Dangerous: too much can cause global warming and climate change.

10 They prevent radiation from being emitted out into space.

11 Miller and Urey demonstrated that hydrogen, ammonia, methane and water vapour can react together to produce amino acids, which form the building blocks of proteins.

2 Using materials our planet provides

▶ 2.1

1 4

▶ 2.2

1 A carbon atom has 6 protons, 6 electrons and 6 neutrons.

▶ 2.3

1 2 atoms in O=O, 2 atoms in N≡N, 3 atoms in O=C=O, 5 atoms in CH_4

▶ 2.4

1 Plastics/polymers, fuels

▶ 2.5

1 Carbon and carbon monoxide

▶ 2.6

1 Both processes rely on the boiling points of the different materials being different. With crude oil, the process happens above $0\,°C$, with air it is below $0\,°C$.

▶ 2.7

1 Dust, noise and toxic chemicals may affect local area. There will be an initial rise in employment when the mine opens, but unemployment when it closes.

▶ 2.8

1 No atoms have been lost of gained. They are only rearranged.

2 Reactants: iron oxide, carbon monoxide Products: iron, carbon dioxide

3 $458 - 44 = 414$ tonnes

▶ 2.9

1 Nitrogen, hydrogen, oxygen

▶ 2.10

1 Don't waste materials, use energy efficiently to keep energy costs low, keep transport costs low by choosing location carefully, control worker wages.

Answers to end of chapter questions

1 Compounds: CH_3, KI, HBr, $AgNO_3$; Elements: O, Cl, H_2, Br_2, Ca

2 Positive

3 They have different boiling points.

4 carbon + oxygen $\longrightarrow$ carbon dioxide carbon + carbon dioxide $\longrightarrow$ carbon monoxide carbon monoxide + iron oxide $\longrightarrow$ carbon dioxide + iron

5 Aluminium is more reactive than carbon.

6 Helium – cooling and balloons; argon – lighting, lasers; nitrogen – fertilisers, making ammonia, freezing agent, food preservative

7 **a** Dust and toxic chemicals in soil/water
   b Increased carbon dioxide emissions

8 Consider the needs of stakeholders, reduce the environmental damage. (Higher tier – consider phytomining.)

9 Their atoms are rearranged, which changes them into the products.

10 **a** X + Y
   b 23 + 32 = 55, so X + Y = 55. X = 12, so Y = 43 g

11 A: $Mg + H_2SO_4 \longrightarrow MgSO_4 + H_2$ – balanced
B: $Mg + HCl \longrightarrow MgCl_2 + H_2$ – not balanced, there should be 2HCl

12 Price of reactants/raw materials can increase, workers' wages can increase, taxes can increase, cost of energy can increase, and raw materials can be wasted.

Unit 1, Theme 2

3 Life on our planet

▶ 3.1

1 Sorting things into groups based on similar features

2 They produce their own food through photosynthesis.

▶ 3.2

1 How different species have evolved from a common ancestor

2 How species depend on each other

▶ 3.3

1 A place where an organism lives

2 The habitat and all the organisms living there

3 The living organisms present in a habitat

▶ 3.4

1 A characteristic that enables an organism to live successfully in its habitat

2 Large surface area to volume ratio to maximise heat loss; thick fur on top of body for shade; thin fur everywhere else to maximise heat loss.

▶ 3.5

1 The process through which organisms have developed from a common ancestor

2 Organisms with the characteristics most suited to their environment survive and reproduce, passing on these characteristics to their offspring.

▶ 3.6

1 A plant hormone

2 Phototropism means growing towards light, gravitropism means growing towards gravity.

3 They make parts of a plant grow faster than others.

Answers to end of chapter questions

1 A place where an organism lives

2 A microorganism that can live in an extreme environment

3 Water, sunlight and nutrients

4 Three of: food, shelter, mates and a suitable territory

5 Gravitropism means growing towards gravity. When the roots grow towards gravity they gain anchorage, and normally end up closer to a water source.

6 A community is all the living organisms present in an environment, whereas an ecosystem refers to all the living organisms and non-living factors present in a habitat.

7 Any three from: large feet avoid sinking into snow and hairs on the soles of the feet help to grip, thick layers of fat and fur which provide insulation, small surface area to volume ratio to reduce heat loss, white fur to act as camouflage, sharp teeth and claws make good weapons to catch and eat prey.

To name and identify species (it makes it easier to find out which species an organism belongs to if everything is organised), to predict characteristics (if several members in a group have a particular characteristic another species in the group may have the characteristic) and to find evolutionary links (species in the same group probably share characteristics because they have evolved from a common ancestor.

Auxin is destroyed by sunlight. Therefore normal-sized plant cells are found on the side of the plant facing the Sun. More auxin is present on the shaded side of the plant. This causes the cells to lengthen on this side of the plant resulting in the plant growing towards the light.

10 Organisms in a species show a wide range of variation (caused by genetic differences). The organisms with the characteristics that are most suited to the environment are most likely to survive and reproduce: 'survival of the fittest'. Genes from successful organisms are passed to the offspring in the next generation. This process is then repeated many times.

4 Biomass, energy flow and the importance of carbon

4.1
All the living and non-living matter present in an area

Producers make their own food by photosynthesis. Consumers have to eat other organisms to gain energy.

4.2
Photosynthesis

10%

4.3
The numbers decrease and the size increases.

Pyramids of biomass take into account the size and number of organisms present, pyramids of number only take into account the number of organisms present.

4.4
A microorganism which breaks down dead organic material into soluble nutrients

Small animals which speed up decomposition by breaking down dead organic material into very small pieces

4.5
Through photosynthesis

Through respiration, decomposition and combustion

4.6
Bones and shells

Answers to end of chapter questions

1 A producer makes its own food by the process of photosynthesis. A consumer has to eat another organism to gain energy.

2 Any three from: fossil fuels, rocks, oceans and animals and plants (temporary store)

3 a Green plants
b Beetles and grasshoppers

4 Decomposers break down dead organic material releasing soluble nutrients. Detritivores shred organic material into very small pieces making it easier for decomposers to break it down.

5 a When a tree is the producer
b Pyramids of numbers only take into account the number of organisms present in a food chain, not their size.

6 Any three from: not all parts of a plant or animal may be eaten, not all parts of the plant or animal can be digested and it will be lost from the food chain in faeces, energy released by respiration is used for movement and other body processes and energy is lost from the body in faeces and urine (waste products).

7 Plants obtain the nutrients needed for growth from the soil. These are then passed on to animals when the plant is eaten. When plants and animals die, decomposers release the nutrients trapped in them back into the soil, where they are absorbed by plants and the cycle begins again.

8 More fossil fuels have to be burned releasing carbon dioxide, also trees may have to be cut down for fuel. This adds carbon dioxide to the atmosphere when they are burned and reduces the amount of carbon dioxide which is removed from the atmosphere through photosynthesis.

9 a $120\,kJ - 75\,kJ - 35\,kJ = 10\,kJ$
b $10/120 \times 100 = 8.3\%$

Answers to examination-style questions

1 a The wavelength is stretched/The light from stars appears more red/Red-shift is happening *(1 mark)*
Because stars are moving away *(1 mark)*
b It is expanding *(1 mark)*

2 a Marks awarded for this answer will be determined by the Quality of Written Communication (QWC) as well as the standard of the scientific response.
There is a clear, balanced and detailed description of what happens to the Earth's plates to cause earthquakes. The answer shows almost faultless spelling, punctuation and grammar. It is coherent and in an organised, logical sequence. It contains a range of appropriate or relevant specialist terms used accurately including: convection current, density, mantle, plate boundaries. *(5–6 marks)*
There is a description of what happens to the Earth's plates to cause earthquakes. There are some errors in spelling, punctuation and grammar. The answer has some structure and organisation.
The use of specialist terms has been attempted, but not always accurately. *(3–4 marks)*
There is a brief description of what happens to the Earth's plates to cause earthquakes which has little clarity and detail. The spelling, punctuation and grammar are very weak. The answer is poorly organised with almost no specialist terms and/or their use demonstrating a general lack of understanding of their meaning. *(1–2 marks)*
No relevant comments. *(0 marks)*
Some examples of points made in the response:
● The plates move because...
  – The core of the Earth is very hot.
  – Convection currents are set up because the mantle contains some molten rock.
  – The mantle nearest the core heats up and becomes less dense so it rises.
  – The heated mantle as it moves away from the core cools and becomes less dense so it falls.
  – Colder mantle moves in closer to the core to take the place of the heated mantle that moves away.

  – This causes convection currents to occur inside the mantle.
● When the plates are pushed together they cause earthquakes on the plate boundaries.
b Any **two** from:
  Volcanic activity released gases that formed the early atmosphere
  Including methane/ammonia/carbon dioxide
  Release of water vapour that condensed to form oceans *(2 marks)*

3 a Plants use it for photosynthesis. *(1 mark)*
b Keep temperatures on Earth warm and stable
Stable temperatures support life/Water is not frozen. *(2 marks)*

4 a Otter = 1
Small fish = 5
Insect larvae = 200
Water weeds = 20 (2 marks for all 4 correct, 1 mark for 2 correct) *(2 marks)*
b Dry them.
Weigh them. *(2 marks)*
c Lost due to:
  Respiration
  Urine and faeces
  The otter does not eat all the parts of the small fish *(3 marks)*

5 a i Plants need water for photosynthesis. *(1 mark)*
  ii The plants that are most suited to keeping their water are more likely to survive and breed successfully.
  This means that the genes for thinner leaves are passed on to the next generation.
  This is known as natural selection. *(3 marks)*
b Deep/extensive root system
so that the plant collects lots of water. *(2 marks)*

6 a Population of pale speckled moth decreased
Because it was more easily seen on polluted tree bark
And so more pale moths were eaten by predators. *(3 marks)*
b Pale and speckled has higher population now
Because the Clean Air Act in 1956 means that there will be less pollution
So more moss grown on trees
Creating more camouflage for the paler moth/less for the dark moth
So fewer pale speckled moths/more dark moths are eaten by predators. *(5 marks)*

7 a Fractional distillation [Allow 1 mark.]
Air is first cooled so that it becomes a liquid/Air is condensed.
Then it is heated/allowed to warm up.
The different gases are collected from different points in the distillation column.
Nitrogen with the lowest boiling point rises to the top and is collected as the gas.
Oxygen is collected from the bottom as a liquid.
(Allow 3 marks for all the points, allow 2 marks for 3 of the points and 1 mark for 2 of the points.) *(4 marks)*
b Used to make ammonia/used for making fertilisers *(1 mark)*

8 a Iron oxide and coke are heated.
Coke burns to produce carbon dioxide.
The carbon dioxide reacts with the coke to produce carbon monoxide.
Iron oxide is reduced producing molten iron. *(4 marks)*
b Reactants ($Fe_2O_3 + CO$)
Products ($Fe + CO_2$)
Balanced ($Fe_2O_3 + 3CO \longrightarrow 2Fe + 3CO_2$) *(3 marks)*

Unit 2, Theme 1

5 Body systems

➤ 5.1
1 Cells that detect a stimulus – a change in the environment.
2 Any three from: light, sound, smell, taste, touch, and heat
3 A reflex action occurs without thinking – as the reaction doesn't involve the brain it occurs faster.

➤ 5.2
1 A longitudinal wave
2 Between 20 Hz and 20 000 Hz.

➤ 5.3
1 Chemical messengers that are made in glands

➤ 5.4
1 Insulin and glucagon
2 Diabetes

➤ 5.5
1 37 °C.
2 Water, salt and urea
3 To trap a layer of air – insulation

➤ 5.6
1 Hydrochloric acid/bile

➤ 5.7
1 Positive

➤ 5.8
1 A hydroxide ion has a –1 charge and a hydrogen ion has a +1 charge; overall, –1 + 1 = 0

Answers to end of chapter questions

1 The maintenance of a constant internal environment
2 Sensory, relay and motor neurons
3 Hairs on the skin lie flat, they begin to sweat and blood vessels supplying capillaries near the surface of their skin widen.
4 A controlled reaction involves the brain and therefore is slower than a reflex action which doesn't involve the brain.
5 When cymbals are crashed together they vibrate. This causes the air particles next to them to vibrate. These vibrations pass energy onto neighbouring particles. This eventually causes your ear drum to vibrate – and so you hear the sound.
6 a When blood glucose levels are too high insulin causes the liver to remove glucose from the blood and store it as glycogen. This reduces blood glucose levels.
  b Glucagon causes the liver to turn stored glycogen back into glucose. The glucose is then released into the blood stream.
7 The body maintains a steady state through the process of negative feedback. This means that any changes which affect the body are reversed, and returned to normal.
8 Chemical burns, heartburn/indigestion/ulcers, tooth decay
9 So it can neutralise the acid that causes tooth decay
10 Hydrochloric acid, to kill bacteria and help digest food
11 Weak alkalis used to neutralise excess stomach acid (indigestion, heartburn)
12 a hydrochloric acid + magnesium hydroxide $\longrightarrow$ water + magnesium chloride
  b $2HCl + Mg(OH)_2 \longrightarrow 2H_2O + MgCl_2$

6 Human inheritance and genetic disorders

➤ 6.1
1 It contains the information that determines what a cell will look like, and what it does.
2 A section of DNA that codes for one characteristic

➤ 6.2
1 The differences between individuals in a species
2 Any two from: height, weight, intelligence, sporting ability, etc.

➤ 6.3
1 Dominant alleles will always be expressed. Recessive alleles will only be expressed if both the genes in a pair are recessive.
2 Four

➤ 6.4–6.5
1 If there is a high risk of having a child with a genetically inherited disorder
2 Because they only have one copy of the 'faulty' allele
3 The presence of extra digits on hands or feet
4 The X chromosome
5 Cells that can develop into any cell type in the body

Answers to end of chapter questions

1 a The cell membrane
  b In the cytoplasm
  c The nucleus
2 Differences between species
3 A dominant allele will always be expressed, whereas a recessive allele will only be expressed if there are two copies of the recessive allele.
4 Gene, chromosome, DNA, nucleus, cell
5 Genetic variation: blood group, eye colour
  Environmental variation: none
  Both: height, weight, intelligence, sporting ability
6 a ss
  b Ss
  c They will have no symptoms of the disease.
7 a

	p	p
P	Pp	Pp
p	pp	pp

  b 50%

Unit 2, Theme 2

7 Materials used to construct our homes

➤ 7.1
1 Floors, walls, windows

➤ 7.2
1 Carbon dioxide

➤ 7.3
1 They both contain sand, cement and water, but concrete also contains small stones.

➤ 7.4
1 It is ductile and a good electrical conductor.

➤ 7.5
1 propene $\longrightarrow$ polypropene

➤ 7.6
1 They have a high melting point and are good thermal insulators.
2 Glass reinforced plastic, because it's light and strong.

➤ 7.7
1 Wood absorbs carbon dioxide from the atmosphere as it grows.
2 Straw, clay and soil
3 Advantage: smaller carbon footprint; disadvantage: shorter lifespan

Answers to end of chapter questions

1 Heated in a rotary kiln
2 quicklime: calcium carbonate $\longrightarrow$ calcium oxide + carbon dioxide; $CaCO_3 \longrightarrow CaO + CO_2$
  slaked lime: calcium oxide + water $\longrightarrow$ calcium hydroxide; $CaO + H_2O \longrightarrow Ca(OH)_2$
3 Small stones
4 Sand and sodium carbonate
5 E.g. reinforced concrete walls, wiring, piping, guttering, flashing, window frames
6 $n(C_2H_4) \longrightarrow (C_2H_4)_n$
7 Ceramics aren't attacked by cleaning products.
8 A composite
9 Advantage: smaller carbon footprint; disadvantage: doesn't last as long as brick.
10 Cob homes
11 Straw absorbs carbon dioxide from the atmosphere when growing.
12 Cob absorbs heat during the day, keeping the home cool. It releases heat during the night, heating the home, i.e. passive solar heating.

8 Fuels and electricity

➤ 8.1
1 It might freeze in mid-air.

➤ 8.2
1 Complete combustion takes place in plenty of oxygen and produces carbon dioxide and water; incomplete combustion takes place in limited oxygen and also produces carbon monoxide and soot (carbon) and releases less energy.

➤ 8.3
1 $(12 \times 2) + 2 = 26$
2 $CH_4 + 2O_2 \longrightarrow CO_2 + 2H_2O$

➤ 8.4
1 Running out/Cannot be replaced at rate they are used up
2 Cars produce sulfur dioxide and nitrogen oxides. These gases can cause acid rain.

➤ 8.5
1 The stages are: burning coal, heating water, producing steam, turning turbines, spinning generators, generating electricity.

➤ 8.6
1 The stages are: nuclear fission reactions, heating water, producing steam, turning turbines, spinning generators, generating electricity.
2 Advantage: e.g. no greenhouse gases emitted. Disadvantage: e.g. radioactive waste must be stored and disposed of safely.

➤ 8.7
1 Renewable energy sources will not run out.
2 Some countries have a climate that suits wind, solar or wave power, or have suitable sites for hydroelectricity, tidal or geothermal energy.

8.8

Power station, step-up transformer, power cable, sub-station, step-down transformer, home
High voltages reduce the current in the power cables, which reduces energy wastage by heating as the electricity is distributed.
Scientists believe that magnetic fields from high-voltage cables do not cause harm because they are much weaker than the Earth's magnetic field, and experiments using magnetic fields on cells have not harmed them.

Answers to end of chapter questions

1 Carbon dioxide and water
2 hexane + oxygen $\longrightarrow$ water + carbon dioxide
3 $C_3H_8 + 5O_2 \longrightarrow 3CO_2 + 4H_2O$
4 Soot (carbon) and carbon monoxide
5 C_3H_8, $C_{11}H_{24}$, C_9H_{20}, C_8H_{18}
6 Acid rain, climate change, oil spills, respiratory problems, carbon monoxide poisoning
7 Hydroelectricity, tidal schemes and waves
8 fission reactions in fuel rod produce heat $\longrightarrow$ heat changes water to steam $\longrightarrow$ steam spins turbines $\longrightarrow$ turbine spins a generator
9 We are using them faster than they can be replaced.
10 Wind turbines only produce energy if it is windy, but hydroelectricity stores water in reservoirs to generate electricity when needed.
11 Hydroelectricity floods large areas; it disrupts river flow.

Unit 2, Theme 3

9 Using energy and radiation

9.1
1 It stays the same.

9.2
1 1.2 kW
2 920 W

9.3
1 0.3 kWh
2 50 Units
3 They use less electricity to provide the same light.

9.4
1 50%

9.5
1 100 m/s
2 10 000 million Hz

9.6
1 a microwaves   b ultraviolet   c infrared

9.7
1 They carry more energy as they have a higher frequency.
2 They can cause cancer if a person is exposed to high doses.

Answers to end of chapter questions

1 joules
2 a electrical energy $\longrightarrow$ kinetic energy + heat and sound energy
   b 50 J
3 Power is the rate of energy transfer.
4 a 50 W
   b 15 000 J or 15 kJ
5 1.2 W
6 75%
7 120 m/s

8 gamma rays, X-rays, ultraviolet, visible light, infrared, microwaves, radio waves
9 They all travel at 300 million m/s through space.
10 Microwaves are used for cooking, for mobile phone networks, and for satellite TV.
11 Gamma rays and X-rays are harmful in large doses.

Answers to examination-style questions

1 a

	f	f
F	Ff	Ff
f	ff	ff

(1 mark for the correct parental genotypes. 1 mark for filling in the Punnett square correctly. 1 mark for circling the correct diseased genotypes (ff)
1 mark for identifying 50%
OR
½
OR
2 out of 4.)   (4 marks)
   b The disease is on a recessive gene.
      So both parents could be carriers of the disease but not suffer from it.   (2 marks)
2 a The blood vessels supplying the skin increase in diameter.
      This allows more blood to flow nearer the surface of the skin.
      This allows the heat from the blood to leave through the surface of the skin.   (3 marks)
   b Water from sweat evaporates from the skin. It takes heat energy with it.   (2 marks)
   c The thermoregulatory centre senses that the body becomes too cold.
      A message is sent to the effectors to switch off any heat loss mechanisms.
      This reverses any changes to the system's steady state so the opposite effect is started and the body then warms up.   (3 marks)
3 a A reaction with hydroxides [accept alkalis or bases] and acids to form salts and other products [accept water].   (2 marks)
   b $CaCO_3$ (1 mark)
   c When limestone is heated/thermal decomposition carbon dioxide is produced.
      Because carbon dioxide is a greenhouse gas, it adds to the effect of global warming/the greenhouse effect.   (2 marks)
   d i $CaO + H_2O \longrightarrow Ca(OH)_2$ (2 marks)
      ii A reaction that releases heat/energy   (1 mark)
4 a A compound containing hydrogen and carbon only   (1 mark)
   b Reactants ($C_3H_8 + O_2$)
      Products ($CO_2 + H_2O$)
      Balancing ($C_3H_8 + 5O_2 \longrightarrow 3CO_2 + 4H_2O$)   (3 marks)
5 a A = chemical, electrical
      B = light
      C = heat   (3 marks)
   b Power = $6 \times 2 = 12$ watts/W   (2 marks)
   c Efficiency = $\dfrac{\text{useful power out}}{\text{total power in}}$
      Efficiency = $\dfrac{9}{12} = 0.75$
      (Allow error carried forward from part b.)   (2 marks)
6 a Copper is cheaper than silver
      But not as good a conductor.   (2 marks)
   b $63.0 \times 10^6 - 45.2 \times 10^6 = 17.8 \times 10^6$
      $17.8 \times 10^6 \div (63.0 \times 10^6) \times 100 = 28.25$ %

Accept working in which 10^6 has been cancelled throughout.   (3 marks)
7 Marks awarded for this answer will be determined by the Quality of Written Communication (QWC) as well as the standard of the scientific response.
There is a clear, balanced and detailed description for and against mobile phones carrying a health warning. The answer shows almost faultless spelling, punctuation and grammar. It is coherent and in an organised, logical sequence. It contains a range of appropriate or relevant specialist terms used accurately including: microwaves, energy, cancer, symptoms.   (5–6 marks)
There is a description for and against mobile phones carrying a health warning. There are some errors in spelling, punctuation and grammar. The answer has some structure and organisation. The use of specialist terms has been attempted, but not always accurately.   (3–4 marks)
There is a brief description for and against mobile phones carrying a health warning which has little clarity and detail. The spelling, punctuation and grammar are very weak. The answer is poorly organised with almost no specialist terms and / or their use demonstrating a general lack of understanding of their meaning.   (1–2 marks)
No relevant comments.   (0 marks)
Some examples of points made in the response:
- Microwaves are used to communicate between phones over long distances
- Microwaves are low-energy electromagnetic waves
- There is very little scientific evidence that these waves damage the body
- Some of the symptoms reported may be due to lifestyle choices
- Mobile phone companies would lose money

Unit 3, Theme 1

10 The use (and misuse) of drugs

10.1
1 Reduce swelling.

10.2
1 Bacteria
2 MRSA

10.3
1 Testing drugs on humans
2 One argument for and one argument against should be stated. Answers could include:

Arguments for testing	Arguments against testing
Shows how drug affects a living body.	Reaction to drug may be different from a human.
Animal lives are not as valued as human lives.	Animals have the right to life.
Animals have a shorter lifecycle so long-term effects can be studied in a relatively short time.	Can cause pain or discomfort for the animal.
Many animals can be tested at one time and it is cheaper than carrying out research on humans.	Many animals die during the testing, or have to be put down after the trial.

10.4

1 They alter chemical reactions in the body.
2 They suffer withdrawal symptoms.

10.5

1 It narrows blood vessels and makes the heart beat faster.
2 Fatty deposits are left on their artery walls reducing blood flow.

10.6

1 It slows down body reactions and can change behaviour.
2 Any three from: stomach ulcers, heart disease, and brain and liver damage (cirrhosis)
3 An alcoholic

Answers to end of chapter questions

1 Medical drugs cure, treat or prevent disease; recreational drugs are purely taken for pleasure.
2 a Antibiotic b Analgesic c Anti-inflammatory/analgesic
3 Antibiotics only kill bacteria.
4 Any two from: alcohol poisoning, increased risk of long-term health problems, antisocial behaviour, increased risk of accidents
5 Stimulants: speed up the nervous system; depressants: slow down the nervous system.
6 C Drug is tested using computer models and human cells.
   B Drug is tested on animals.
   E Drug is tested on a small group of healthy human volunteers.
   F Drug is tested on volunteer patients who have the illness.
   A Drug is tested on a wide range of people.
   D Drug is approved and can be prescribed.
7 a Ethanol
   b It slows down the nervous system.
8 a A person who is dependent on a drug
   b They suffer withdrawal symptoms as the body is no longer being provided with a chemical it has got used to having.
9 Carbon monoxide is a poisonous gas which reduces the blood's oxygen-carrying capacity. Oxygen is transported in blood cells by binding to haemoglobin. But if carbon monoxide is present, it will bind to haemoglobin in preference to oxygen.
10 a Bacteria can spontaneously mutate, and these mutations can lead to some strains becoming resistant to antibiotics.
   b When antibiotics are used, they will kill individual pathogens that do not have antibiotic resistance. However, resistant pathogens will survive. These will then reproduce, increasing the population of the resistant strain.

11 Modern medicine

11.1

1 A pathogen
2 The time taken between a pathogen entering the body and a person feeling unwell
3 Bacteria are larger and have a cell wall. Viruses are smaller and have a protein coat.

11.2

1 Any three from: skin, digestive system, respiratory system, reproductive system and any three from: cook food properly, drink clean sterilised water, maintain good personal hygiene (wash hands), cover cuts

11.3

1 Help the blood to clot
2 Produce antibodies (chemicals) which deactivate microorganisms

11.4

1 They will not suffer from it as they already have antibodies against the disease.
2 Dead or weakened microorganisms

11.5

1 A vaccine has to undergo lots of stages of testing to ensure it is safe before it is given to people.
2 Any three from: fever, sickness, diarrhoea, swollen glands and irritability

11.6

1 Broken bones, tumours and chest infections
2 They monitor radiation dose using film badges and protect themselves by standing behind a lead screen.

11.7

1 Atoms have a neutral charge, ions are charged.
2 a gamma b alpha c beta

11.8

1 Tracers are radioactive substances that are injected into the body or eaten.

Answers to end of chapter questions

1 To clot the blood
2 a Examples include TB, salmonella or pneumonia.
   b Examples include measles, rubella or flu.
3 Any three from: covering your mouth with a handkerchief, using condoms, using sterilised needles or covering cuts with plasters.
4 Produce antibodies which 'fight off' microorganisms.
5 Advantage: protect person from disease. Disadvantage: possible side effects.
6 Immunisation involves a vaccine being inserted into the body. These dead or weakened microorganisms in the vaccine do not make a person ill, but trigger lymphocytes to make antibodies. Antibodies destroy the microorganisms. Some remain in the body to fight off the pathogen quickly if it enters the body again, preventing it causing disease. The person is now immune.
7 X-rays penetrate through soft tissues like skin and muscle. They are absorbed by denser structures, such as bones and teeth. The X-rays that penetrate through the patient expose photographic film. These areas appear black on the developed film. Areas of the film not exposed to X-rays appear white.
8 Ionising radiation comes from the nucleus.
9 Two protons and two neutrons
10 Beta and gamma radiation
11 A gamma camera

Unit 3, Theme 2

12 Improving materials

12.1

1 Anions are negative and are attracted to the positive anode because it has the opposite charge to them.

12.2

1 The cathode

12.3

1 It resists corrosion and looks attractive.

12.4

1 Two of: stents that expand due to body heat to hold blood vessels open, braces to correct teeth that tighten due to body heat, smart packaging for medicines that tells us if they have been stored incorrectly
2 Too expensive. Cheaper to just repaint.

12.5

1 Superconductors have no electrical resistance at very low temperatures.

12.6

1 Chromic materials automatically provide information about their surroundings.

Answers to end of chapter questions

1 a Electrolyte
   b Cathode
   c Anions
   d Electroplating
2 a Pure gold
   b The statue
   c Solution containing gold ions
3 They gain electrons.
4 They lose electrons.
5 UV sensor bracelets to prevent sunburn, rear-view mirrors in cars that darken so the driver isn't dazzled by headlights, spectacles that darken in brighter light.
6 Thermochromic pigments can change colour to tell us if medicine has not been stored at the correct temperature.
7 At room temperature, superconductors have electrical resistance. They have almost zero resistance at very low temperatures.
8 Superconductors are used in MRI scanners and for Maglev trains.
9 Smart materials respond to small changes between different people, or fluctuations in their bodies, e.g. temperature changes.
10 Small scratches repair themselves so the owner does not have to pay for the cost of the repair.

13 Selective breeding and genetic disorders

13.1

1 Choosing the best organisms to breed together to produce offspring with desired characteristics
2 Reduces the gene pool.

13.2

1 Frost-resistant tomato/insect-resistant Bt corn

13.3

1 A circle of DNA found in bacteria
2 Any two from: allows the rapid production of many genetically identical plants, very little space is needed and the plants are disease-free as they are grown in controlled environments.
3 Skin grafts

13.4

1 Advantage (one of): infertile couples may be able to have children, embryos can be selected which do not have a genetically inherited disorder. Disadvantage: designer babies could be created.
2 Advantage: could cure genetically inherited disorders. Disadvantage: could alter a person's appearance or characteristics.

Answers to end of chapter questions

The bacteria produced are genetically identical.
a Highest milk producers
b Highest meat content
c Highest meat content/Hens which lay the most eggs
One of: skin, cartilage
Altering an organism's genes. It is faster and more accurate than selective breeding.
A baby that has been created specifically to have a desired characteristic.
The replacement of a faulty gene (e.g. one that causes a disorder) with a healthy gene in a person's cells.
Any two arguments for genetic engineering from: plants and animals can be produced with desired characteristics such as higher yields, disease and frost resistance, seedless fruits produced, medicines can be produced cheaply and quickly using bacteria, may offer cures for genetically inherited disorders
Any two arguments against genetic engineering from: may reduce an animal's life span or affect an animal's quality of life, if seedless species interact with wild plants it may result in wild plants not producing seeds, techniques could be used to change a person's appearance or characteristics
A tissue sample is scraped from the parent plant, placed on an agar plate containing nutrients and auxins. Samples grow into plantlets which are planted in compost, and grown in a greenhouse where they develop into full-size plants

```
┌─────────────────────────────────────────┐
│ Farmer chooses cow with the highest milk │
│ production and breeds it with his best bull │
└─────────────────────────────────────────┘
                    ↓
┌─────────────────────────────────────────┐
│            Offspring produced            │
└─────────────────────────────────────────┘
                    ↓
┌─────────────────────────────────────────┐
│ Farmer then chooses cow with the highest │
│ milk production and breeds it with his best bull │
└─────────────────────────────────────────┘
                    ↓
┌─────────────────────────────────────────┐
│   This process is repeated for many years │
└─────────────────────────────────────────┘
```

10 1 Plasmids are modified to include the gene for making insulin.
2 The plasmid is then inserted into the bacteria. The bacteria now produce insulin.
3 The bacteria multiply quickly, and produce large quantities of insulin in a fermenter.
4 Bacteria are then killed by heat (sterilisation), leaving behind the insulin.

Unit 3, Theme 3

14 Environmental concerns when making and using products

▐▌▶ 14.1
1 a carbon dioxide, nitrous oxide
  b nitrous oxide, methane

▐▌▶ 14.2
1 The Kyoto Protocol is a legally binding agreement between countries to reduce the amount of greenhouse gases produced.

▐▌▶ 14.3
1 Pesticides kill insects that may damage crops.

2 Indicator species are affected more by changes in pollution levels than other species.

▐▌▶ 14.4
1 Take up less space, can recycle the energy.
2 Polymers
3 Lactic acid

Answers to end of chapter questions

1 Greenhouse gas emissions have increased due to increased travel, electricity generation and intensive farming methods.
2 Denitrification converts nitrates in the soil to nitrous oxide gas.
3 The greenhouse effect is when greenhouse gas molecules absorb and re-radiate radiation from the Earth; global warming is the average increase in global temperatures.
4 Increases in greenhouse gases cause global warming which changes the climate.
5 Since many countries produce greenhouse gas emissions, global emissions will only fall if most countries reduce these levels.
6 Some countries refused to sign the Kyoto Protocol because they did not want to reduce their standard of living.
7 One advantage of artificial fertilisers is increased crop yields; and one disadvantage is eutrophication.
8 If there is sulfur dioxide pollution in a region, the lichen levels on trees will fall.
9 Microbes cannot digest them.
10 Biomass can be used to make polymers that can be digested by microbes.
11 Water soluble polymers degrade faster outdoors. Oxo-degradable polymers degrade in air to allow microbes in. Photo-degradable polymers break down when exposed to light.

15 Our environment at home

▐▌▶ 15.1
1 Carpets are made from wool, which is an insulator, and has trapped air pockets.

▐▌▶ 15.2
1 Draught excluders stop warm air escaping from homes.

▐▌▶ 15.3
1 Black walls in ovens radiate heat well.

▐▌▶ 15.4
1 4 months (1/3 of a year)

▐▌▶ 15.5
1 Walls have a larger surface area so more heat is lost through them compared with windows.

▐▌▶ 15.6
1 Bathroom

▐▌▶ 15.7
1 A rash or chemical burn
2 Hazard symbols are printed on them.

▐▌▶ 15.8
1 It would reduce the efficiency of the boiler.
2 The symptoms (tiredness and headaches) are similar.

▐▌▶ 15.9
1 Radon gas only seeps out of certain types of rocks, which are only found in some parts of the country.

End of chapter answers

1 Conduction, convection, radiation

2 When a liquid is heated from the base or cooled from the top.
3 The particles are not free to move.
4 Light-coloured curtains reduce conduction (trap layers of air), reduce convection (stop draughts), and reflect heat radiation.
5 To reflect radiation and keep the home cool.
6 £10000 ÷ £500 = 20 years
7 It is more cost-effective to install energy saving measures with short payback times.
8 They reduce heat transfers into the home in summer and reduce heat losses from the home in winter.
9 Smokers have already damaged some cells in their lungs and radon is more likely to do even more harm to these cells.
10 Mould can produce spores which cause allergies. Pollen can also cause allergies.
11 Cleaning products
12 Carbon monoxide and carbon/soot
13 It binds to haemoglobin and replaces oxygen in blood.

Answers to examination-style questions

1 a Select a female chicken that lays a lot of eggs and a male chicken whose mother laid a lot of eggs.
    Breed them together.
    This process is repeated over several generations of chicken. (3 marks)
  b It is quicker
    As do not have to wait for several generations. (2 marks)
2 a A: gamma
    B: alpha
    C: beta (3 marks)
  b Used to treat cancer by killing living cells
    OR
    Used for sterilisation as kills microbes
    OR
    As a tracer
    Can be detected outside the body. (2 marks)
  c Alpha
    Because it cannot penetrate the plastic case. (2 marks)
3 a Paracetamol only treats the symptoms.
    Paracetamol does not treat the disease. (2 marks)
  b Flu is caused by a virus. Antibiotics only treat bacterial diseases. (2 marks)
  c So all the microorganisms in the body are killed because if only a few of them survive then they can multiply rapidly and make you ill again. (2 marks)
  d Over-prescribing antibiotics can lead to bacteria which are resistant to the antibiotic. These bacteria are then difficult to treat. (2 marks)
4 Tobacco: lung cancer; reduces oxygen capacity of the blood (2 marks)
  Alcohol: slows down reactions; long-term damage to brain (2 marks)
5 a Marks awarded for this answer will be determined by the Quality of Written Communication (QWC) as well as the standard of the scientific response.
    There is a clear, balanced and detailed description of how a vaccination protects against pathogens. The answer shows almost faultless spelling, punctuation and grammar. It is coherent and in an organised, logical sequence. It contains a range of appropriate

or relevant specialist terms used accurately including: lymphocytes, phagocyte, pathogen, antibody, immune. *(5–6 marks)*
There is a description of how a vaccination protects against pathogens. There are some errors in spelling, punctuation and grammar. The answer has some structure and organisation. The use of specialist terms has been attempted, but not always accurately. *(3–4 marks)*
There is a brief description of how a vaccination protects against pathogens which has little clarity and detail. The spelling, punctuation and grammar are very weak. The answer is poorly organised with almost no specialist terms and/or their use demonstrating a general lack of understanding of their meaning. *(1–2 marks)*
No relevant comments. *(0 marks)*
Some examples of points made in the response:
- A dead, small or weakened form of the disease
- Is injected into the blood stream.
- The lymphocytes make antibodies to stick on the pathogen.

- The phagocytes engulf/eat the pathogen.
- The white blood cells can make the antibodies for that pathogen much quicker next time the disease is encountered.
- The pathogens are destroyed next time before they have chance to have an effect on the body.
- The body is now immune to that disease.

b Platelets
Clot blood/form a barrier *(2 marks)*

6

Chromic material		Aircraft so it doesn't matter if it is scratched
Self-healing paint		Electromagnets
Smart material		Shrink-wrap packaging
Superconductor		Rear-view mirrors

(All correct = 3 marks, 3 correct = 2 marks, 1 or 2 correct = 1 mark.) *(3 marks)*

7 a The process involves the movement of charged particles in an electrolyte.
The article to be plated is attached to/made the cathode. The anode is a bar of the plating metal. The cathode is then placed in a solution containing ions of the plating metal.

During electrolysis, metal is deposited on the article as metal from the anode goes into the solution. *(4 marks)*
b i $M^{n+} + ne^- \longrightarrow M$ *(1 mark)*
ii $M \longrightarrow$ *(1 mark)*
$M^{n+} + ne^-$ *(1 mark)*
c For: Does not rust/corrode OR does not matter if it is scratched.
Against: Poisonous so may kill plants/animals. *(2 marks)*

8 a Combustion of fossil fuels/petrol/diesel
Produces carbon dioxide. *(2 marks)*
b They allow short-wave radiation from the Sun in.
But they do not allow long-wave radiation from the Earth out. *(2 marks)*
c Methane
Produced from landfill sites and agriculture
OR
Nitrous oxide
Produced in power stations and from using nitrogen-based fertilisers *(2 marks)*

The Periodic Table of Elements

1	2												3	4	5	6	7	0
								1 **H** hydrogen 1										4 **He** helium 2
7 **Li** lithium 3	9 **Be** beryllium 4		**Key** relative atomic mass **atomic symbol** name atomic (proton) number										11 **B** boron 5	12 **C** carbon 6	14 **N** nitrogen 7	16 **O** oxygen 8	19 **F** fluorine 9	20 **Ne** neon 10
23 **Na** sodium 11	24 **Mg** magnesium 12												27 **Al** aluminium 13	28 **Si** silicon 14	31 **P** phosphorus 15	32 **S** sulfur 16	35.5 **Cl** chlorine 17	40 **Ar** argon 18
39 **K** potassium 19	40 **Ca** calcium 20	45 **Sc** scandium 21	48 **Ti** titanium 22	51 **V** vanadium 23	52 **Cr** chromium 24	55 **Mn** manganese 25	56 **Fe** iron 26	59 **Co** cobalt 27	59 **Ni** nickel 28	63.5 **Cu** copper 29	65 **Zn** zinc 30	70 **Ga** gallium 31	73 **Ge** germanium 32	75 **As** arsenic 33	79 **Se** selenium 34	80 **Br** bromine 35	84 **Kr** krypton 36
85 **Rb** rubidium 37	88 **Sr** strontium 38	89 **Y** yttrium 39	91 **Zr** zirconium 40	93 **Nb** niobium 41	96 **Mo** molybdenum 42	[98] **Tc** technetium 43	101 **Ru** ruthenium 44	103 **Rh** rhodium 45	106 **Pd** palladium 46	108 **Ag** silver 47	112 **Cd** cadmium 48	115 **In** indium 49	119 **Sn** tin 50	122 **Sb** antimony 51	128 **Te** tellurium 52	127 **I** iodine 53	131 **Xe** xenon 54
133 **Cs** caesium 55	137 **Ba** barium 56	139 **La*** lanthanum 57	178 **Hf** hafnium 72	181 **Ta** tantalum 73	184 **W** tungsten 74	186 **Re** rhenium 75	190 **Os** osmium 76	192 **Ir** iridium 77	195 **Pt** platinum 78	197 **Au** gold 79	201 **Hg** mercury 80	204 **Tl** thallium 81	207 **Pb** lead 82	209 **Bi** bismuth 83	[209] **Po** polonium 84	[210] **At** astatine 85	[222] **Rn** radon 86
[223] **Fr** francium 87	[226] **Ra** radium 88	[227] **Ac*** actinium 89	[261] **Rf** rutherfordium 104	[262] **Db** dubnium 105	[266] **Sg** seaborgium 106	[264] **Bh** bohrium 107	[277] **Hs** hassium 108	[268] **Mt** meitnerium 109	[271] **Ds** darmstadtium 110	[272] **Rg** roentgenium 111		Elements with atomic numbers 112 – 116 have been reported but not fully authenticated						

* The Lanthanides (atomic numbers 58 – 71) and the Actinides (atomic numbers 90 – 103) have been omitted.

Cu and **Cl** have not been rounded to the nearest whole number.